Illuminating Pathways

Distributed by Prioriti AI Media

A Group within Prioriti AI, Inc.

4 W 4th Ave. 6th Floor

San Mateo, CA 94402

For more information, email media@prioriti.ai

First Edition 2024

Hard Cover: ISBN 979-8-218-46011-2

Trade Paperback ISBN 979-8-218-45235-3

eBook ISBN 979-8-3507-3326-6

Audience

Note the glossary at the end of the book to provide the reader with the necessary terminology used throughout the book. The table of contents is your north star for this book as it is written more so like a collection of key topics for someone to quickly go to and consume as opposed to a cover-to-cover read. Think of a playbook or textbook. It can be read sequentially, or the reader can jump around from chapter to chapter, topic to topic.

Contents

FOREWORD

The Business Intelligence (BI) software category is tried, tested and proven and includes companies such as Business Objects, Microstrategy, Cognos, Tableau, and others. BI software has enabled companies to analyze data captured in traditional "systems of record" [e.g., Enterprise Resource Planning (ERP), Customer Relationship Management (CRM), Supply Chain Management (SCM), et al] and offers executives and managers a historical view of the business, using visual aids such as charts, graphs and reports.

BI software can deliver tangible hindsight, which is useful when analyzing various elements of the business retrospectively. However, these analytical tools do not provide such insights on their own. Humans are required to examine the data contained in the depths of the charts, graphs and reports and left to identify any new markets, products, and business opportunities or remedy any operational issues that may lurk within.

Executives have longed for a "holy grail": the ability to interact with their business systems and from those interactions generate a multitude of future business scenarios, product opportunities, and operational issues. Armed with these predictive insights, they have dreamed of confidently prioritizing and strategically planning the future of the business and, when necessary, taking corrective action in time to potentially avoid a business calamity.

This has largely remained a business and technological pipedream…until now.

Recent advances in computer science collectively labeled –albeit somewhat erroneously– Artificial Intelligence (AI), are quickly turning what was once a pipedream into reality. New capabilities such as Generative AI, Retrieval Augmented Generation, Generative Adversarial Networks (GANs) and Variational Autoencoders (VAEs), all part of the AI "technology ecosystem" are giving rise to a new software category – Decision Intelligence- and several subcategories such as: Enterprise Decision Intelligence and Enterprise Strategy Intelligence.

Decision Intelligence applications are built on a new technology stack (e.g., Large Language Models - LLMs). They are designed from the ground up using AI to operate in real-time as a "system of intelligence" combining the data in exogenous systems (e.g., multi-modal LLMs, current stock market data, published government and business statistics, blog posts, et al) with data from internal systems of record to produce novel ideas and insights – without human intervention. Decision Intelligence applications can accept informal text, verbal and digital inputs – "prompts" – and create a variety of outputs (e.g., text, graphs, charts, video, audio) to provide decision-makers with unique insights and potential competitive advantage.

Skip's book "Illuminating Pathways" is your literary sherpa and will guide you on a journey of enlightenment so that you can better understand, implement, and exploit Decision Intelligence applications. Skip introduces myriad AI and DI concepts, explains what they are, what they do, and how you can leverage them. He includes case studies showing how to apply DI in the Insurance, Healthcare, Pharmaceutical, ESG, Financial Services, CPG and Retail industries. He dives into the various Key Performance Indicators (KPIs) for Decision Intelligence applications; you can use these KPIs to measure the effectiveness of your own DI systems.

It's not hyperbolic to claim that Decision Intelligence applications will deliver a step function increase in GDP by dramatically improving the performance of all businesses, in all industries. As a result, "Illuminating Pathways" is required reading for any business executive who wants their company to remain viable.

Bruce Cleveland
Technology Executive, Venture Capitalist and Author

Introduction

In today's fast-paced and data-driven business landscape, organizations face a myriad of complex decisions that can significantly impact their success. Making informed decisions requires a deep understanding of data, patterns, and trends. This is where decision intelligence comes into play. Decision intelligence combines data analytics, artificial intelligence (AI), and human expertise to illuminate the pathways towards optimal decision-making. And within the realm of decision intelligence, generative AI emerges as a powerful tool, capable of generating new content, simulating scenarios, and providing valuable insights.

We embark on a journey to explore the convergence of decision intelligence and generative AI within the enterprise ecosystem. We will delve into the transformative potential of generative AI in illuminating pathways for making strategic decisions, fostering innovation, and driving business success. By leveraging the power of generative AI, organizations can gain a competitive edge and navigate the complexities of the modern business landscape with confidence.

We will uncover the ways in which generative AI can simulate multiple scenarios, assess risks, and enable decision-makers to explore different pathways before making crucial choices. The data-driven insights and pattern recognition capabilities of generative AI will be explored, demonstrating how organizations can uncover hidden patterns, anticipate trends, and gain a deeper understanding of their customers and markets. Additionally, we will showcase how generative AI acts as a catalyst for idea generation and innovation, sparking creativity and revolutionizing traditional approaches. Furthermore, we will examine the role of generative AI in personalization and customer-centric decision-making, where organizations can tailor products, services, and experiences to meet the unique needs and preferences of individual customers. The article will also shed light on how generative AI enhances design processes, automates creative content generation, and accelerates rapid prototyping, enabling organizations to stay at the forefront of innovation and deliver exceptional experiences.

Throughout this exploration, we will emphasize the importance of ethical considerations, privacy protection, and fairness in the deployment of generative AI within the enterprise ecosystem. Ensuring responsible and accountable practices will guide organizations towards making ethical decisions, avoiding biases, and safeguarding the interests of their stakeholders.

Join us on this journey as we unravel the potential of generative AI in decision intelligence, offering insights, strategies, and real-world examples that illuminate pathways for success. Together, we will navigate the intricate enterprise ecosystem, harness the transformative power of generative AI, and make informed decisions that propel organizations towards a future of innovation and growth.

Figure 1

"The true sign of intelligence is not knowledge but imagination." Albert Einstein

Applying Imagination Instead of Knowledge to Decision Intelligence

Unleashing the Power of Creative Thinking

When we think of decision-making, knowledge and data are often seen as the primary drivers of informed choices. We rely on facts, figures, and historical information to guide our decisions. However, there is another powerful aspect of human cognition that often goes underutilized in the realm of decision intelligence: imagination. By incorporating imagination into the decision-making process, we can unlock new perspectives, innovative solutions, and unprecedented opportunities.

Imagination is a uniquely human ability that enables us to mentally simulate possibilities, envision outcomes, and generate creative ideas. It goes beyond what we already know and allows us to explore uncharted territories. By tapping into our imagination, we can break free from the confines of conventional thinking and embark on a journey of discovery.

When applied to decision intelligence, imagination can play a transformative role. It can help us think beyond existing data and knowledge, enabling us to consider scenarios that have not yet been experienced or documented. By imagining alternative futures, we can identify potential risks and opportunities that may not be apparent through traditional analytical approaches.

Imagination also encourages us to challenge assumptions and question established norms. It pushes us to explore unconventional solutions and consider out-of-the-box strategies. This approach is particularly valuable in complex and ambiguous decision contexts where traditional knowledge-based approaches may fall short. Imagination allows us to connect disparate ideas, draw upon diverse perspectives, and create innovative solutions that can drive transformative change.

To apply imagination to decision intelligence, it is essential to foster a culture that encourages and values creative thinking. Organizations can create dedicated spaces for brainstorming and ideation, where individuals are encouraged to share imaginative ideas without fear of judgment. Embracing diversity and fostering interdisciplinary collaboration can also fuel imagination by bringing together individuals with different backgrounds, experiences, and perspectives.

In addition to fostering a supportive environment, decision intelligence tools can be enhanced to incorporate imaginative thinking. AI algorithms can be designed not only to analyze historical data but also to simulate and generate future scenarios. These tools can help decision-makers explore various "what-if" scenarios, simulate the potential outcomes of different choices, and identify hidden patterns or trends.

Furthermore, integrating imagination into the decision-making process can be facilitated through techniques such as storytelling, visualization, and scenario planning. By creating narratives and visual representations of potential futures, decision-makers can better understand the implications of their choices and explore the consequences from different angles. These techniques not only stimulate imagination but also facilitate communication and consensus-building among stakeholders.

It is important to note that imagination should not replace knowledge or data-driven analysis. Rather, it should complement and enhance these approaches. Knowledge provides the foundation upon which imaginative thinking can thrive. The combination of knowledge, data, and imagination empowers decision-makers to make more holistic and innovative choices.

In conclusion, applying imagination to decision intelligence opens up a world of possibilities. By unleashing our creative thinking, we can transcend the boundaries of traditional knowledge-based decision-making and explore new frontiers. Embracing imagination in the decision-making process allows us to envision futures that have not yet materialized and create pathways to turn those visions into reality. By harnessing the power of imagination, we can revolutionize decision intelligence and unlock unprecedented opportunities for growth and success.

Strategy

Figure 2

Chapter 1: The Significance of Decision Intelligence in Enterprise Strategy

In the dynamic and competitive landscape of the business world, strategic decision-making plays a pivotal role in the success and growth of enterprises. However, the complexity and uncertainty surrounding strategic choices can often overwhelm decision-makers, leading to suboptimal outcomes. This is where the concept of decision intelligence steps in to transform the way organizations approach strategic decision-making. Decision intelligence combines data-driven insights, analytical tools, and human expertise to guide organizations towards informed and effective strategies.

In this section, we will delve into the significance of decision intelligence in enterprise strategy and explore how it empowers organizations to make better decisions.

1. Unleashing Data-Driven Insights:

One of the primary benefits of generative AI is its ability to unlock valuable insights hidden within vast amounts of data. By leveraging generative models such as generative adversarial networks (GANs) or variational autoencoders (VAEs), organizations can generate synthetic data, augment existing datasets, and uncover patterns, trends, and relationships that may not be apparent through traditional analysis. This empowers decision-makers with a more comprehensive understanding of the business landscape, enabling them to make data-driven decisions based on a broader set of insights.

2. Enhancing Scenario Analysis and Prediction:

Generative AI enables organizations to simulate and analyze multiple scenarios, improving their ability to predict potential outcomes. By training generative models on historical data, organizations can generate synthetic scenarios that represent different possibilities. This allows decision-makers to explore the potential impact of various strategies, assess risks, and make more informed choices. Generative AI fosters a culture of proactive decision-making by providing decision-makers with the tools to anticipate challenges, adapt to changing circumstances, and optimize outcomes.

3. Fostering Creativity and Innovation:

Generative AI has the power to unleash creativity and drive innovation within organizations. By training generative models on diverse datasets, organizations can leverage the creative capabilities of AI algorithms to generate novel ideas and explore alternative solutions. This opens up new avenues for problem-solving and ideation, enabling decision-makers to challenge conventional thinking and uncover innovative strategies. Generative AI acts as a catalyst for human creativity, sparking new perspectives and driving breakthroughs in decision-making processes.

4. Personalization and Customer-Centricity:

In an era of increasing customer expectations, generative AI allows organizations to deliver personalized experiences and drive customer-centric decision-making. By analyzing customer data and generating personalized recommendations, generative AI algorithms enable organizations to understand individual preferences and tailor their strategies, products, and services accordingly. This enhances customer satisfaction, loyalty, and engagement, resulting in better decision outcomes and improved business performance.

5. Optimizing Resource Allocation and Efficiency:

Generative AI empowers organizations to optimize resource allocation and drive operational efficiency. By analyzing patterns, historical data, and complex relationships, generative AI algorithms can provide valuable insights into resource utilization. This helps decision-makers make informed choices about resource allocation, identify bottlenecks, and optimize workflows. By leveraging generative AI, organizations can streamline processes, reduce costs, and maximize the utilization of financial, human, and technological resources.

6. Mitigating Risks and Uncertainty:

Decision-making often involves inherent risks and uncertainties. Generative AI can assist in mitigating these risks by providing decision-makers with a deeper understanding of potential challenges and their potential impact. Through scenario analysis, generative AI algorithms can simulate risk scenarios, evaluate their likelihood and consequences, and help organizations develop robust risk mitigation strategies. This empowers decision-makers to make proactive and risk-informed decisions, reducing the likelihood of adverse outcomes.

Conclusion:

Generative AI presents a multitude of opportunities for organizations to improve their decision-making processes. By embracing generative AI, organizations can unlock data-driven insights, enhance scenario analysis and prediction, foster creativity and innovation, drive personalization and customer-centricity, optimize resource allocation, and mitigate risks and uncertainties. As organizations harness the potential of generative AI, they gain a competitive advantage by making more informed, proactive decisions that drive innovation, efficiency, and business success. The era of improved decision-making through generative AI has arrived, and organizations that embrace these opportunities will lead the way in their respective industries.

Addressing Generative AI Concerns, Pitfalls, Lack of Human Interaction

The Pitfalls of Enterprise Strategy Intelligence Generative AI Without Human Intervention

Introduction: Enterprise Strategy Intelligence (ESI) has evolved significantly with the advent of generative AI technologies. These AI systems have the ability to generate strategies, insights, and recommendations based on vast amounts of data and complex algorithms. While the potential benefits of using generative AI in ESI are undeniable, it is crucial to acknowledge the pitfalls that may arise when relying solely on AI without human intervention. In this section, we will explore the risks and limitations of using generative AI in ESI without adequate human oversight and intervention.

The Lack of Contextual Understanding: One of the key challenges with generative AI in ESI is its limited ability to comprehend and interpret contextual nuances. While AI algorithms excel at processing large amounts of data and identifying patterns, they often struggle to understand the intricacies of real-world business dynamics, market conditions, and social factors. Without human intervention, the generated strategies may lack the necessary contextual understanding, leading to flawed or impractical recommendations that do not align with the organization's goals or the broader business landscape.

Bias and Ethical Considerations: Generative AI systems are trained on historical data, which may contain inherent biases or reflect historical inequities. Without human intervention, these biases can persist or even amplify in the generated strategies. The lack of ethical considerations in AI models can result in unintended consequences, such as discriminatory practices, unequal resource allocation, or biased decision-making. Human intervention is essential to identify and address biases, ensuring that the generated strategies adhere to ethical standards and promote fairness and inclusivity.

Limited Creativity and Innovation: While AI algorithms excel at analyzing data and identifying patterns, they often lack the capacity for creativity and innovation. ESI requires strategic thinking, imagination, and a deep understanding of market dynamics. Human intervention is crucial in injecting fresh perspectives, challenging assumptions, and thinking beyond the boundaries of data-driven insights. Without human involvement, generative AI may produce strategies that are overly conservative or fail to identify novel opportunities for growth and competitive advantage.

Inability to Account for Complex Intangible Factors: ESI involves analyzing both quantitative and qualitative factors, including customer sentiment, brand reputation, and organizational culture. While generative AI can handle structured data, it struggles to capture and interpret unstructured data or intangible factors that are vital in shaping strategy. Human intervention is necessary to assess subjective information, make qualitative judgments, and consider the broader organizational context. Ignoring these complex intangible factors can result in strategies that fail to resonate with stakeholders or overlook critical aspects of the business.

Legal and Regulatory Compliance: The use of generative AI in ESI must comply with legal and regulatory frameworks, particularly when dealing with sensitive data or making strategic decisions with significant consequences. Without human oversight, AI systems may inadvertently violate data privacy laws, misinterpret regulatory guidelines, or make decisions that expose the organization to legal risks. Human intervention ensures the compliance of generated strategies with applicable laws and regulations, safeguarding the organization's reputation and minimizing legal liabilities.

Conclusion: Generative AI has the potential to revolutionize Enterprise Strategy Intelligence, offering powerful insights and recommendations. However, it is essential to recognize the pitfalls and limitations of relying solely on AI without human intervention. To mitigate the risks, organizations must strike a balance by combining the strengths of AI technologies with human expertise and judgment. By integrating human intervention into the ESI process, organizations can leverage the power of generative AI while ensuring contextual understanding, addressing biases, fostering creativity, accounting for intangible factors, and complying with legal and regulatory requirements. The collaboration between humans and AI is key to unlocking the true potential of Enterprise Strategy Intelligence in driving sustainable growth and competitive advantage.

Navigating the Concerns of Generative AI: Harnessing the Power Responsibly

Introduction: Generative Artificial Intelligence (AI) has garnered significant attention and excitement for its ability to create content, generate insights, and drive innovation. However, as this technology evolves, it is crucial to address the concerns that arise alongside its immense potential. In this section, we will delve into the concerns surrounding generative AI and explore the importance of responsible implementation to ensure ethical and beneficial outcomes.

1. **Ethical Considerations:** Generative AI systems are trained on vast amounts of data, raising concerns about privacy, bias, and fairness. It is essential to carefully select and curate training data to minimize biases and ensure inclusivity. Additionally, organizations must prioritize transparency, accountability, and consent when deploying generative AI to maintain ethical standards and foster trust among stakeholders.

2. **Misinformation and Manipulation:** Generative AI has the capacity to create highly realistic and convincing content, which opens the door to potential misuse and manipulation. The dissemination of false information or deepfake content can have severe social, political, and economic implications. Organizations and individuals must be vigilant in verifying the authenticity and credibility of content generated by AI systems, employing fact-checking mechanisms, and promoting media literacy to combat misinformation.

3. **Lack of Contextual Understanding:** Generative AI often lacks the ability to comprehend nuanced contextual factors that humans readily understand. It can struggle to recognize subtle cultural nuances, historical contexts, or specific domain knowledge. As a result, there is a risk of generating content or insights that may be misleading, irrelevant, or out of touch with real-world dynamics. Human oversight and intervention are crucial

to ensuring that generative AI is used in conjunction with human expertise to provide a holistic understanding of the context.

4. **Intellectual Property and Copyright Concerns:** Generative AI systems have the potential to create content that may infringe upon intellectual property rights or copyright laws. Organizations utilizing generative AI must understand and respect intellectual property boundaries, ensuring compliance with legal frameworks and obtaining necessary permissions when generating content based on existing works. Proper attribution and adherence to copyright laws are essential to avoid legal ramifications.

5. **Dependency and Job Displacement:** The rapid advancement of generative AI has led to concerns about job displacement and dependency on AI systems. While AI can automate certain tasks, organizations must focus on reskilling and upskilling employees to adapt to evolving roles and leverage the technology effectively. Responsible implementation of generative AI involves striking a balance between automation and human collaboration to maximize productivity and create new opportunities.

Conclusion: Generative AI holds immense potential for innovation and creativity, but it is crucial to address the concerns surrounding its use. Responsible implementation of generative AI involves proactive measures to address ethical considerations, combat misinformation, ensure contextual understanding, respect intellectual property rights, and mitigate job displacement. By combining the power of generative AI with human oversight and intervention, organizations can harness this technology responsibly and unlock its true potential for positive impact. As we navigate the evolving landscape of generative AI, an ethical and responsible approach will pave the way for a future that balances technological advancement with the well-being of society at large.

Exploring the Limitations of Generative AI in Enterprise Strategy

Introduction: Generative Artificial Intelligence (AI) has gained considerable attention for its potential to transform various industries, including enterprise strategy. By leveraging vast amounts of data and advanced algorithms, generative AI systems can generate insights, recommendations, and even strategies. However, it is crucial to understand the limitations of generative AI in the context of enterprise strategy. In this section, we will explore the key constraints and challenges that organizations should be aware of when utilizing generative AI for strategic decision-making.

1. **Lack of Contextual Understanding:** One of the primary limitations of generative AI is its inability to fully grasp the contextual nuances that influence enterprise strategy.

While AI algorithms excel at processing and analyzing data, they often struggle to comprehend the intricate interplay of social, economic, and cultural factors that shape strategic decisions. Contextual understanding requires human judgment, experience, and domain expertise, which generative AI may not possess. Therefore, relying solely on generative AI for enterprise strategy may lead to recommendations that do not adequately account for the complex and dynamic nature of real-world business environments.

2. **Interpretation and Bias:** Generative AI systems are trained on existing data sets, which can introduce biases and limitations in the generated output. These biases might reflect historical inequities or the limitations of the training data itself. Without proper intervention, generative AI can perpetuate biases or produce recommendations that are skewed or incomplete. Human involvement is necessary to interpret and evaluate the generated output, ensuring that it aligns with ethical and strategic considerations. Moreover, organizations must be mindful of the potential ethical implications of AI-generated strategies, especially when it comes to social impact and fairness.

3. **Creativity and Innovation:** While generative AI can excel at identifying patterns and generating insights based on existing data, it may struggle to generate truly innovative and creative strategies. Innovation often requires thinking beyond the boundaries of available data and exploring unconventional approaches. Human creativity, imagination, and intuition play a vital role in fostering innovative thinking and identifying new opportunities. Relying solely on generative AI might limit the organization's ability to think outside the box and explore disruptive strategies that can drive competitive advantage.

4. **Data Limitations and Quality:** The effectiveness of generative AI heavily depends on the quality and availability of data. In enterprise strategy, the data required to generate meaningful insights and strategies may be limited or difficult to acquire. Additionally, incomplete or inaccurate data can lead to flawed recommendations or misguided decisions. Organizations must ensure that the data used to train generative AI models is relevant, reliable, and comprehensive. It is also crucial to regularly evaluate and update the training data to account for changing business dynamics and market trends.

5. **Legal and Regulatory Compliance:** Deploying generative AI for enterprise strategy must adhere to legal and regulatory frameworks. AI-generated strategies should comply with industry-specific regulations, privacy laws, and intellectual property rights. However, generative AI systems may lack the ability to fully understand and interpret complex legal requirements. Human oversight is essential to ensure compliance and

mitigate the risks of legal and regulatory violations. Organizations must maintain a close partnership between legal experts and AI specialists to align AI-generated strategies with legal and ethical considerations.

Conclusion: Generative AI offers significant potential in the realm of enterprise strategy, but it is essential to acknowledge its limitations. While generative AI can augment decision-making processes by providing data-driven insights, it should not replace human judgment, context understanding, creativity, and ethical considerations. Organizations must leverage generative AI as a tool that complements human expertise, allowing for a collaborative approach to enterprise strategy. By understanding and navigating the limitations of generative AI, organizations can harness its strengths while maximizing the strategic value derived from human intelligence and experience.

Figure 3

Chapter 2: Unleashing the Power of Generative AI in Decision Intelligence

In the realm of decision intelligence, where organizations strive to make informed and impactful choices, generative AI is emerging as a powerful tool. Generative AI, a subset of artificial intelligence, leverages advanced algorithms and models to generate new and valuable insights. By combining the power of data-driven analysis and algorithmic creativity, generative AI is revolutionizing decision-making processes and enhancing decision intelligence. In this section, we will explore the power of generative AI in decision intelligence and how it empowers organizations to unlock new possibilities.

1. Data Synthesis and Augmentation:

One of the key strengths of generative AI is its ability to synthesize and augment data. Decision intelligence relies on data-driven insights, and generative AI enhances this process by generating synthetic data that closely mimics real-world scenarios. By training generative models such as generative adversarial networks (GANs) or variational autoencoders (VAEs), organizations can generate new data points, enrich existing datasets, and explore a wider range of possibilities. This synthesis and augmentation of data enable decision-makers to have a more comprehensive and accurate understanding of the factors influencing their decisions.

2. Scenario Generation and Analysis:

Generative AI allows organizations to generate and analyze multiple scenarios, enabling a deeper understanding of potential outcomes. By training generative models on historical data, organizations can generate synthetic scenarios that represent various possibilities. These scenarios can then be analyzed, allowing decision-makers to assess the potential risks, opportunities, and implications associated with different courses of action. The ability to generate and analyze scenarios using generative AI enhances decision intelligence by providing a more comprehensive view of the potential outcomes and enabling proactive decision-making.

3. Creative Ideation and Problem-Solving:

Decision intelligence goes beyond data analysis; it also requires creative ideation and problem-solving. Generative AI brings a new dimension to decision intelligence by fostering creativity and enabling innovative thinking. By training generative models on diverse datasets, organizations can leverage the creativity of AI algorithms to generate new ideas, challenge conventional thinking, and explore alternative solutions. This creative ideation powered by generative AI can lead to breakthroughs, novel strategies, and innovative approaches to complex problems.

4. Personalization and Customer Insights:

Generative AI enables organizations to leverage customer data and deliver personalized experiences. By analyzing vast amounts of customer data and generating personalized recommendations, generative AI algorithms can uncover individual preferences and provide tailored strategies, products, or services. This level of personalization enhances decision intelligence by enabling organizations to make data-driven decisions that are highly relevant and appealing to their target customers. It allows organizations to gain deeper insights into customer behavior, improve customer satisfaction, and drive better business outcomes.

5. Optimization and Resource Allocation:

Decision intelligence involves optimizing resource allocation to maximize efficiency and achieve desired outcomes. Generative AI aids in this process by analyzing patterns, historical data, and complex relationships to optimize resource allocation. By training generative models on relevant data, organizations can gain valuable insights into resource utilization, identify areas of improvement, and optimize workflows. This enables decision-makers to allocate resources more effectively, improve operational efficiency, and achieve strategic objectives with greater precision.

6. Risk Assessment and Mitigation:

Generative AI plays a crucial role in risk assessment and mitigation. Decision intelligence requires an understanding of potential risks and their potential impact on decision outcomes. Generative AI facilitates risk assessment by generating synthetic risk scenarios based on historical data and analyzing their likelihood and consequences. This empowers decision-makers to identify and mitigate risks proactively, develop robust risk mitigation strategies, and make informed decisions that factor in potential risks and uncertainties.

Conclusion:

Generative AI is revolutionizing decision intelligence by unleashing new capabilities and possibilities. Through data synthesis and augmentation, scenario generation and analysis, creative ideation, personalization, resource optimization, and risk assessment, generative AI empowers organizations to make more informed and impactful decisions. By harnessing the power of generative AI, organizations can unlock new insights, foster innovation, personalize customer experiences, optimize resource allocation, and mitigate risks. The integration of generative AI into decision intelligence heralds a new era of data-driven, creative, and proactive decision-making, positioning organizations for success in the ever-evolving business landscape.

Demystifying Generative AI and Its Relevance to Decision Intelligence

In recent years, the field of artificial intelligence has witnessed remarkable advancements, with generative AI emerging as a powerful and transformative technology. Generative AI, a subset of AI, focuses on the creation and synthesis of new data, images, or text using complex algorithms and models. This innovative approach has the potential to revolutionize decision intelligence by providing organizations with new insights, creative solutions, and enhanced decision-making capabilities. In this section, we will demystify generative AI and explore its relevance to decision intelligence.

What is Generative AI?

Generative AI refers to a class of AI algorithms and models that learn patterns and structures from existing data to generate new, original content. These algorithms, such as generative adversarial networks (GANs) and variational autoencoders (VAEs), can produce realistic images, text, or data points that resemble the patterns and characteristics of the training data. Generative AI algorithms learn from vast amounts of data and can generate new content that is indistinguishable from what a human might create.

The Role of Generative AI in Decision Intelligence:

Decision intelligence involves making informed choices based on data, analysis, and expert knowledge. Generative AI enhances decision intelligence in several ways:

1. **Data Synthesis and Augmentation:** Generative AI allows organizations to synthesize new data points that closely resemble the patterns and characteristics of real-world data. By generating synthetic data, decision-makers can augment existing datasets, address data scarcity issues, and explore a wider range of possibilities. This synthesis and augmentation of data provide decision-makers with a more comprehensive and accurate understanding of the factors influencing their decisions.

2. **Scenario Generation and Analysis:** Generative AI enables the creation of synthetic scenarios based on historical data, allowing decision-makers to simulate and analyze multiple potential outcomes. These generated scenarios help decision-makers assess the risks, opportunities, and implications associated with different strategic choices. By exploring various scenarios, decision-makers can make more informed decisions and develop robust strategies that account for potential challenges and uncertainties.

3. **Creative Ideation and Problem-Solving:** Generative AI fosters creativity by generating new and innovative ideas. By training generative models on diverse datasets, organizations can leverage the creative capabilities of AI algorithms to challenge conventional thinking, explore alternative solutions, and drive creative problem-solving. This infusion of generative AI in decision intelligence empowers decision-makers to discover novel approaches and breakthrough solutions.

4. **Personalization and Customer Insights:** Generative AI helps organizations deliver personalized experiences to customers. By analyzing vast amounts of customer data, generative AI algorithms can generate personalized recommendations and tailor strategies, products, or services to individual preferences. This personalization enhances decision intelligence by enabling organizations to make data-driven decisions that cater to the unique needs and preferences of their customers.

5. **Optimization and Resource Allocation:** Generative AI contributes to optimizing resource allocation by analyzing patterns, historical data, and complex relationships. By training generative models on relevant data, organizations gain valuable insights into resource utilization, identify areas of improvement, and optimize workflows. This optimization of resource allocation enables decision-makers to allocate resources more

effectively, improve operational efficiency, and maximize outcomes

Conclusion:

Generative AI brings a new dimension to decision intelligence by providing organizations with enhanced capabilities to synthesize data, generate scenarios, foster creativity, personalize experiences, and optimize resource allocation. By leveraging generative AI, organizations can unlock new insights, drive innovation, and make more informed and impactful decisions. As the potential of generative AI continues to unfold, decision intelligence stands to benefit from this transformative technology, enabling organizations to navigate complexity, drive growth, and achieve strategic success.

Exploring the Frontier of Generative AI: Retrieval-Augmented Generation

Introduction:

In recent years, artificial intelligence (AI) has made remarkable strides in natural language processing, particularly in the domain of generative models. One fascinating area within this field is retrieval-augmented generation, a technique that combines the strengths of both generative models and retrieval systems. This innovative approach has opened up new avenues for creating more contextually relevant and coherent text, with applications ranging from content creation to dialogue systems and beyond.

Understanding Generative AI:

Generative AI refers to a class of models designed to generate new content, such as text, images, or music, that is indistinguishable from human-created content. These models, often based on deep learning architectures like recurrent neural networks (RNNs) or transformers, learn to mimic the patterns and structures present in the data they are trained on. They achieve this by probabilistically generating sequences of tokens, with the goal of producing coherent and contextually relevant outputs.

Challenges in Generative Models:

While generative models have shown impressive capabilities, they also face inherent challenges. One major issue is maintaining coherence and relevance in generated output, especially when dealing with long-form text or complex tasks that require a deep understanding of context. Traditional generative models may struggle with this, often producing nonsensical or off-topic content.

Enter Retrieval-Augmented Generation:

Retrieval-augmented generation seeks to address these challenges by integrating a retrieval mechanism into the generative process. In this approach, the model has access to a database or knowledge base containing relevant information that it can retrieve and incorporate into the generation process. By leveraging external knowledge, the model can produce more contextually grounded and coherent outputs.

How Retrieval-Augmented Generation Works:

At its core, retrieval-augmented generation involves two key components: a generative model and a retrieval mechanism. The generative model, typically a transformer-based architecture like OpenAI's GPT (Generative Pre-trained Transformer), generates the initial output based on the provided prompt. Meanwhile, the retrieval mechanism retrieves relevant information from a knowledge base using the prompt as a query. This retrieved information is then fused with the generated output to produce a final coherent response.

Benefits and Applications:

Retrieval-augmented generation offers several advantages over traditional generative models. By incorporating external knowledge, these models can produce more contextually relevant and coherent text, making them well-suited for tasks such as content creation, question answering, and dialogue systems. In addition, retrieval-augmented generation can help mitigate issues like bias and misinformation by ensuring that generated content is grounded in factual information from trusted sources.

Moreover, these models can adapt to various domains and tasks by leveraging domain-specific knowledge bases. For example, a retrieval-augmented generation model trained on medical literature could assist healthcare professionals in generating accurate and relevant medical reports or providing personalized treatment recommendations.

Challenges and Future Directions:

Despite its promise, retrieval-augmented generation also presents challenges and limitations. One key concern is the scalability and efficiency of the retrieval mechanism, especially when dealing with large knowledge bases or real-time applications. Additionally, ensuring the reliability and accuracy of the retrieved information is crucial, as errors or inaccuracies could propagate into the generated output.

Looking ahead, future research in retrieval-augmented generation will likely focus on addressing these challenges while exploring novel techniques to enhance the performance and capabilities of these models. This may involve advancements in retrieval algorithms, improved integration with generative architectures, and the development of more sophisticated evaluation metrics to assess the quality of generated output.

Conclusion:

Retrieval-augmented generation represents an exciting frontier in the field of generative AI, offering a powerful approach to creating contextually relevant and coherent text. By combining the strengths of generative models with external knowledge sources, these models have the potential to revolutionize various applications, from content generation to knowledge dissemination and beyond. As research in this area continues to advance, we can expect to see even more sophisticated and capable AI systems that push the boundaries of what is possible in natural language understanding and generation.

Exploring Generative Models and their Applications in Strategic Assessments

In the ever-evolving landscape of strategic assessments, organizations are constantly seeking innovative approaches to gain a competitive edge. Generative models, a branch of artificial intelligence (AI), have emerged as powerful tools that can revolutionize the way strategic assessments are conducted. These models, such as generative adversarial networks (GANs) and variational autoencoders (VAEs), offer the capability to generate new data based on existing patterns, enabling organizations to gain valuable insights and make more informed strategic decisions. In this section, we will explore generative models and their applications in strategic assessments.

Understanding Generative Models:

Generative models are AI algorithms that learn from a given dataset to generate new, realistic data that resembles the original training data. They capture the underlying patterns and structures in the data and use them to create new instances that are statistically similar to the training set. Generative models enable organizations to generate synthetic data, images, or text, allowing for enhanced analysis, exploration, and decision-making.

Applications of Generative Models in Strategic Assessments

1. Data Augmentation and Enrichment:

Generative models are invaluable in data augmentation, where synthetic data is generated to supplement existing datasets. By generating new instances of data that closely resemble the original data distribution, generative models address data scarcity issues and enhance the robustness of strategic assessments. This augmented dataset allows for more comprehensive analysis, improves the accuracy of statistical models, and mitigates biases resulting from limited data.

2. Scenario Generation and Analysis:

Generative models play a crucial role in generating synthetic scenarios for strategic assessments. By training generative models on historical data, organizations can generate new scenarios that represent a range of possibilities. These synthetic scenarios provide decision-makers with a broader understanding of potential outcomes and allow them to assess the risks, uncertainties, and opportunities associated with different strategies. This helps in exploring alternative pathways, optimizing decision-making, and developing resilient strategies.

3. Sensitivity Analysis and Risk Assessment:

Generative models are instrumental in sensitivity analysis, enabling organizations to evaluate the impact of changing variables or parameters on strategic assessments. By manipulating specific features or inputs in the generative model, decision-makers can observe how different factors affect the outcomes. This helps identify critical variables, assess their influence on strategic assessments, and gain insights into potential risks or vulnerabilities.

4. Ideation and Creative Thinking:

Generative models foster creativity and innovative thinking in strategic assessments. By training generative models on diverse datasets, organizations can explore new possibilities and generate novel ideas. This encourages decision-makers to think outside the box, challenge conventional thinking, and develop innovative strategies. Generative models can generate a range of options, inspiring creative solutions and driving breakthroughs in strategic decision-making.

5. Simulation and Prediction:

Generative models can simulate new data points based on learned patterns, allowing organizations to forecast and predict future trends. By generating synthetic data, decision-makers can simulate various scenarios and assess their potential outcomes. This enables them to make more informed decisions, optimize resource allocation, and anticipate market dynamics. Generative models enhance the accuracy and reliability of predictions, providing a valuable tool for strategic assessments.

Conclusion:

Generative models are transforming the landscape of strategic assessments, offering organizations novel ways to gain insights, explore scenarios, and make informed decisions. Through data augmentation, scenario generation and analysis, sensitivity analysis, ideation, and simulation, generative models provide decision-makers with a powerful toolkit to enhance strategic assessments. As organizations harness the potential of generative models, they unlock new possibilities, improve the accuracy of strategic assessments, and position themselves for success in an ever-changing business environment.

The Potential of Generative Adversarial Networks (GANs) and Variational Autoencoders (VAEs)

Artificial intelligence (AI) has witnessed significant advancements in recent years, particularly in the realm of generative models. Among them, Generative Adversarial Networks (GANs) and Variational Autoencoders (VAEs) have emerged as powerful techniques for generating new, realistic data. These models have revolutionized the fields of image synthesis, data augmentation, and creative content generation. In this section, we will explore the potential of GANs and VAEs and their applications across various domains.

Understanding Generative Adversarial Networks (GANs):

GANs consist of two key components: a generator and a discriminator. The generator takes random noise as input and generates synthetic data samples. The discriminator, on the other hand, aims to distinguish between the real and synthetic data. The two components are trained simultaneously in a competitive manner, with the generator improving its ability to generate realistic data as the discriminator becomes more adept at distinguishing between real and synthetic samples. This adversarial training process leads to the generation of high-quality, realistic data that closely resembles the training dataset.

Applications of GANs:

1. **Image Synthesis and Augmentation:** GANs have demonstrated exceptional capabilities in generating realistic images. They can synthesize new images by learning from a dataset and capturing the underlying patterns and structures. This has applications in various domains, including art, fashion, and computer graphics. GANs can also be

used for data augmentation, generating additional samples that expand the diversity of a dataset, improving the performance of machine learning models, and mitigating overfitting.

2. **Style Transfer and Image Editing:** GANs can be used for style transfer, where the style of one image is applied to another. This enables the transformation of images into different artistic styles or altering specific attributes, such as the color palette or texture. GANs also facilitate image editing, allowing users to modify specific features or characteristics of an image while maintaining its overall coherence.

3. **Text-to-Image Synthesis:** GANs have been employed in text-to-image synthesis tasks, generating images from textual descriptions. This has applications in areas such as creative content generation, visual storytelling, and virtual reality. GANs can convert text descriptions into visual representations, enabling the creation of personalized visual content based on textual input.

Understanding Variational Autoencoders (VAEs):

VAEs are generative models that learn the latent space representation of data. Unlike GANs, VAEs utilize an encoder-decoder architecture. The encoder compresses the input data into a lower-dimensional latent space, while the decoder reconstructs the original input from the latent representation. VAEs focus on learning the underlying distribution of the data, allowing for the generation of new data samples from the learned latent space.

Applications of VAEs:

1. **Data Generation and Imputation:** VAEs excel at generating new data samples that resemble the training dataset. This has applications in data synthesis and completion, particularly in situations where data is missing or incomplete. VAEs can fill in the missing values or generate new instances that adhere to the data distribution, enhancing the completeness and quality of datasets.

2. **Anomaly Detection and Outlier Identification:** VAEs can learn the normal patterns and structures of a dataset. By reconstructing input data, VAEs can identify anomalies or outliers that deviate significantly from the learned data distribution. This makes VAEs valuable for tasks such as fraud detection, cybersecurity, and quality control.

3. **Dimensionality Reduction and Visualization:** VAEs enable the compression of high-dimensional data into a lower-dimensional latent space. This can facilitate dimensionality reduction, aiding in visualizing and exploring complex datasets. VAEs can capture the essential features and structure of the data, simplifying the analysis and interpretation of large datasets.

Conclusion:

Generative Adversarial Networks (GANs) and Variational Autoencoders (VAEs) have revolutionized the field of generative modeling and have far-reaching applications across diverse domains. GANs excel in image synthesis, style transfer, and content generation, while VAEs offer capabilities in data generation, anomaly detection, and dimensionality reduction. As these techniques continue to evolve, their potential for creative content generation, data augmentation, anomaly detection, and advanced analytics will pave the way for exciting new developments in AI and its practical applications.

Figure 4

Chapter 3: Assessing Strategy Initiatives with Generative AI

Assessing Strategy Initiatives with Generative AI: Unlocking New Possibilities

In today's fast-paced and complex business landscape, organizations face the challenge of assessing and prioritizing strategy initiatives effectively. The emergence of generative AI has brought about a paradigm shift in the way organizations evaluate and assess their strategic initiatives. Generative AI, a subset of artificial intelligence, leverages advanced algorithms and models to generate new insights, simulate scenarios, and optimize decision-making. In this section, we will explore how organizations can harness the power of generative AI to assess their strategy initiatives and unlock new possibilities.

1. Data Augmentation and Synthesis:

Generative AI enables organizations to augment and synthesize data, addressing the common challenge of limited or incomplete data in strategy assessments. By training generative models on existing data, organizations can generate synthetic data points that closely resemble real-world scenarios. This augmented data provides decision-makers with a broader and more comprehensive dataset, enabling them to assess the potential impact and outcomes of their strategy initiatives with greater accuracy.

2. Scenario Simulation and Analysis:

Generative AI empowers organizations to simulate and analyze multiple scenarios related to their strategy initiatives. By generating synthetic scenarios based on historical data, generative models allow decision-makers to explore a wide range of possibilities and assess the potential risks and opportunities associated with different strategic choices. This simulation and analysis enable organizations to make informed decisions, anticipate potential challenges, and optimize their strategy initiatives for success.

3. Risk Assessment and Mitigation:

Assessing and mitigating risks is a crucial aspect of strategy initiatives. Generative AI enhances the risk assessment process by generating synthetic risk scenarios based on historical data and analyzing their potential impact. Decision-makers can evaluate different risk scenarios, assess their likelihood and consequences, and develop effective risk mitigation strategies. Generative AI empowers organizations to proactively identify and address potential risks, ensuring that strategy initiatives are aligned with risk tolerance and business objectives.

4. Creative Ideation and Innovation:

Generative AI fosters creative thinking and innovation in strategy assessments. By training generative models on diverse datasets, organizations can leverage the creativity of AI algorithms to explore alternative solutions and generate new ideas. This infusion of generative AI into strategy assessments opens up new avenues for problem-solving, challenges conventional thinking, and drives innovation. Decision-makers can tap into the power of generative AI to unlock fresh perspectives and develop unique strategies that propel their organizations forward.

5. Resource Optimization and Allocation:

Generative AI supports organizations in optimizing resource allocation for their strategy initiatives. By analyzing patterns, historical data, and complex relationships, generative models provide insights into resource utilization. Decision-makers can identify bottlenecks, optimize resource allocation, and allocate resources strategically to initiatives that have the highest potential for success. Generative AI helps organizations make data-driven decisions, maximize the efficiency of resource utilization, and achieve their strategic goals effectively.

6. Decision Support and Prioritization:

Generative AI serves as a valuable decision support tool for prioritizing strategy initiatives. By leveraging generative models, organizations can evaluate the potential impact, feasibility, and alignment of different initiatives with their strategic objectives. Decision-makers can prioritize initiatives based on objective assessment criteria, such as generated synthetic data, simulated scenarios, and risk analysis. Generative AI facilitates a structured and data-driven approach to prioritizing strategy initiatives, ensuring that resources are allocated to the most promising and impactful initiatives.

Conclusion:

Generative AI has emerged as a powerful tool for assessing strategy initiatives, revolutionizing the decision-making process for organizations. By augmenting data, simulating scenarios, mitigating risks, fostering innovation, optimizing resource allocation, and providing decision support, generative AI empowers organizations to make informed and strategic choices. As organizations embrace generative AI in their strategy assessments, they unlock new possibilities, gain a competitive advantage, and set the stage for success in an increasingly dynamic and uncertain business environment.

Leveraging Generative AI to Analyze and Evaluate Strategy Initiatives

In the ever-changing business landscape, organizations are constantly seeking innovative ways to analyze and evaluate their strategy initiatives effectively. The rise of generative AI has opened up new possibilities, enabling organizations to harness the power of advanced algorithms and models to gain deeper insights, simulate scenarios, and optimize decision-making. In this section, we will explore how organizations can leverage generative AI to analyze and evaluate their strategy initiatives, leading to more informed and impactful decision-making.

1. Generating Synthetic Data:

Generative AI enables organizations to generate synthetic data that closely resembles real-world scenarios. By training generative models on existing data, organizations can create new data points that expand the dataset, allowing for a more comprehensive analysis of strategy initiatives. This synthetic data generation facilitates a more robust evaluation, providing decision-makers with a broader view of the potential outcomes and enabling them to make data-driven decisions based on a larger set of insights.

2. Simulating and Analyzing Scenarios:

Generative AI allows organizations to simulate and analyze multiple scenarios related to their strategy initiatives. By generating synthetic scenarios based on historical data, generative models enable decision-makers to explore the potential impact and outcomes of different strategic choices. This simulation and analysis provide valuable insights into the risks, opportunities, and trade-offs associated with each scenario, empowering decision-makers to make informed decisions and optimize their strategy initiatives accordingly.

3. Assessing Risk and Uncertainty:

Generative AI plays a crucial role in assessing and managing risk and uncertainty in strategy initiatives. By generating synthetic risk scenarios, organizations can evaluate the potential risks and their potential impact on the outcomes of their initiatives. This allows decision-makers to proactively identify and mitigate risks, develop contingency plans, and allocate resources effectively. Generative AI enhances risk assessment by providing a more comprehensive and dynamic view of potential risks, enabling organizations to make informed decisions in the face of uncertainty.

4. Optimizing Resource Allocation:

Generative AI aids in optimizing resource allocation for strategy initiatives. By analyzing patterns, historical data, and complex relationships, generative models provide insights into resource utilization and performance. This enables decision-makers to allocate resources strategically, identifying areas where resources can be optimized or reallocated to maximize the impact of their strategy initiatives. Generative AI facilitates a data-driven approach to resource allocation, ensuring that resources are utilized efficiently and effectively.

5. Creative Ideation and Innovation:

Generative AI fosters creative ideation and innovation in the evaluation of strategy initiatives. By training generative models on diverse datasets, organizations can leverage the creative capabilities of AI algorithms to generate new ideas and explore alternative solutions. This infusion of generative AI encourages decision-makers to think outside the box, challenge conventional thinking, and develop innovative strategies. Generative AI acts as a catalyst for creativity, sparking new perspectives and driving breakthroughs in the evaluation and optimization of strategy initiatives.

6. Data-Driven Decision-Making:

Generative AI enables data-driven decision-making in the evaluation of strategy initiatives. By generating synthetic data, simulating scenarios, and assessing risk and resource allocation, generative models provide decision-makers with a solid foundation of insights and evidence. This facilitates objective and informed decision-making, reducing reliance on intuition or subjective assessments. Generative AI empowers organizations to make strategic choices based on comprehensive analysis and evaluation, increasing the likelihood of success for their initiatives.

Conclusion:

Generative AI offers organizations a powerful toolset for analyzing and evaluating strategy initiatives. By generating synthetic data, simulating scenarios, assessing risk, optimizing resource allocation, fostering creativity, and enabling data-driven decision-making, generative AI enhances the evaluation process and empowers organizations to make informed and impactful decisions. As organizations embrace generative AI, they gain a competitive advantage by unlocking deeper insights, optimizing their strategy initiatives, and driving success in a rapidly changing business environment.

Uncovering Hidden Patterns And Relationships Using Generative Models

In the vast sea of data that organizations collect and analyze, there are often hidden patterns and relationships that go unnoticed. These elusive insights can hold the key to unlocking new opportunities, optimizing processes, and making informed decisions. Generative models, a branch of artificial intelligence (AI), have emerged as powerful tools to uncover these hidden patterns and relationships. By leveraging advanced algorithms and training techniques, generative models have the capacity to reveal valuable insights that may have otherwise remained hidden. In this section, we will explore how generative models can uncover hidden patterns and relationships, and the impact they can have on various domains.

Understanding Generative Models:

Generative models are AI algorithms that learn the underlying distribution of a given dataset and generate new instances that closely resemble the original data. Two popular types of generative models are generative adversarial networks (GANs) and variational autoencoders (VAEs). GANs consist of a generator and a discriminator, while VAEs focus on learning the latent representation of the data. Both models excel at capturing the intricate patterns and structures within the data, enabling the generation of synthetic samples that reflect the original data distribution.

Unveiling Hidden Patterns:

Generative models have the unique capability to uncover hidden patterns within datasets, even when these patterns are not explicitly defined or understood. By training on large volumes of data, generative models can capture complex relationships and dependencies that may not be apparent to human observers. These models can generate synthetic data points that exhibit these hidden patterns, allowing decision-makers to gain new insights and a deeper understanding of the underlying dynamics at play.

1. Image and Video Analysis:

Generative models have had a significant impact on image and video analysis, revealing hidden patterns and structures within visual data. By training on extensive datasets, generative models can generate synthetic images or videos that exhibit consistent patterns, such as the representation of specific objects, textures, or styles. These generated samples can unveil latent attributes or visual relationships that were previously unknown. This knowledge can be leveraged for tasks like content generation, style transfer, or anomaly detection.

2. Natural Language Processing:

Generative models have proven invaluable in the field of natural language processing (NLP). By training on vast amounts of text data, generative models can generate coherent and contextually relevant sentences or paragraphs. This ability to generate text allows these models to unveil hidden patterns within the language, such as sentence structures, sentiment, or semantic relationships. Generative models in NLP have applications in text generation, language translation, and dialogue systems.

3. Financial Data Analysis:

In the domain of finance, generative models have been used to uncover hidden patterns and relationships within complex financial datasets. By training on historical financial data, generative models can generate synthetic financial scenarios that reflect market dynamics and interdependencies. These generated scenarios can help uncover hidden correlations, detect anomalies, or simulate what-if scenarios. Such insights can be utilized for risk assessment, portfolio optimization, or fraud detection.

4. Healthcare and Biomedical Research:

Generative models are revolutionizing healthcare and biomedical research by revealing hidden patterns within patient data, genomics, and medical images. By training on diverse datasets, generative models can generate synthetic medical images or patient profiles that capture underlying disease patterns or genetic variations. These generated samples can assist in diagnosis, drug discovery, or treatment planning, enabling researchers and healthcare professionals to make more informed decisions.

5. Customer Behavior Analysis:

Generative models are also useful in understanding and analyzing customer behavior. By training on customer data, such as purchase history or browsing patterns, generative models can generate synthetic customer profiles that reveal hidden preferences, purchase patterns, or demographic relationships. This knowledge can help businesses personalize marketing strategies, improve customer segmentation, or identify cross-selling opportunities.

Conclusion:

Generative models have the potential to uncover hidden patterns and relationships across various domains, from image analysis to financial data, from healthcare to customer behavior analysis. By leveraging their ability to capture complex dependencies and generate synthetic samples, generative models empower organizations to gain deeper insights and make data-driven decisions. As generative models continue to advance, they hold the promise of unlocking new opportunities, optimizing processes, and driving innovation in a wide range of industries.

Enhancing Data Synthesis and Simulation for Comprehensive Assessments

In the realm of data-driven decision-making, comprehensive assessments are crucial for organizations to gain insights, make informed choices, and drive successful outcomes. However, traditional assessment approaches often face limitations when it comes to data availability, complexity, and the ability to simulate diverse scenarios. This is where advancements in data synthesis and simulation techniques, empowered by artificial intelligence (AI) and machine learning, have the potential to revolutionize the assessment process. By enhancing data synthesis and simulation, organizations can achieve more comprehensive assessments that uncover hidden insights, enable scenario exploration, and drive better decision-making. In this section, we will explore how organizations can leverage data synthesis and simulation techniques to enhance their assessments.

1. Data Synthesis for Enhanced Analysis:

Data synthesis techniques, fueled by AI algorithms like generative models, enable organizations to generate synthetic data that closely resembles real-world datasets. This data synthesis process addresses data scarcity issues and expands the available data for analysis. By synthesizing additional data points, organizations can enhance their assessments by incorporating a more diverse and comprehensive dataset. This allows for a more accurate analysis of patterns, trends, and relationships within the data, leading to more informed decision-making.

2. Simulation of Scenarios:

Simulation techniques play a vital role in comprehensive assessments by allowing organizations to explore various scenarios and assess their potential outcomes. By leveraging AI-powered simulation models, organizations can simulate different scenarios based on synthesized data and real-world parameters. These simulations help decision-makers understand the potential risks, opportunities, and consequences associated with different strategic choices. By simulating multiple scenarios, organizations gain a deeper understanding of the complex dynamics at play, enabling them to make more informed and robust decisions.

3. Incorporating Uncertainty and Risk Analysis:

Comprehensive assessments should take into account uncertainties and potential risks. Data synthesis and simulation techniques allow organizations to incorporate uncertainty factors into their assessments. By generating synthetic data points that represent uncertainties or variations, organizations can simulate the impact of these factors on the outcomes. This helps decision-makers assess the potential risks, evaluate mitigation strategies, and make more risk-aware decisions. Enhanced data synthesis and simulation enable organizations to capture the intricacies of uncertainties and risk factors, providing a more comprehensive assessment of their initiatives.

4. Complex System Modeling:

Many strategic initiatives involve complex systems with numerous interdependencies and feedback loops. Enhancing data synthesis and simulation techniques enables organizations to model these complex systems more accurately. By synthesizing data that reflects the intricate relationships within the system, organizations can simulate the behavior of the system under different conditions. This allows decision-makers to assess the impact of their initiatives on the overall system, identify potential bottlenecks or unintended consequences, and optimize their strategies accordingly.

5. Optimizing Resource Allocation:

Data synthesis and simulation techniques also aid in optimizing resource allocation for strategic initiatives. By generating synthetic data points and simulating scenarios, organizations can evaluate the resource requirements and the potential outcomes of different resource allocation strategies. This enables decision-makers to allocate resources more effectively, identify areas where resources can be optimized, and ensure that resources are aligned with the most impactful initiatives. Enhanced data synthesis and simulation facilitate data-driven resource allocation, maximizing the return on investment and driving better overall performance.

6. Mitigating Bias and Assumptions:

Data synthesis and simulation techniques provide an opportunity to mitigate biases and assumptions that may be inherent in the available data. By generating synthetic data points, organizations can diversify the dataset and reduce the influence of any specific biases or assumptions present in the original data. This allows for a more objective and unbiased assessment, enabling decision-makers to make fairer and more equitable choices.

Conclusion:

Enhancing data synthesis and simulation techniques opens up new possibilities for comprehensive assessments. By leveraging AI algorithms, organizations can generate synthetic data, simulate diverse scenarios, incorporate uncertainties and risks, model complex systems, optimize resource allocation, and mitigate biases. These enhancements empower decision-makers to gain deeper insights, explore a wider range of possibilities, and make more informed decisions. As organizations embrace data synthesis and simulation techniques, they can drive better outcomes, optimize their strategies, and achieve success in a rapidly changing business environment.

Incorporating Multidimensional Perspectives to Capture a Holistic View

In the realm of decision-making, it is crucial to capture a holistic view that encompasses various dimensions and perspectives. Traditional approaches often focus on specific aspects, which may lead to incomplete or biased assessments. However, organizations are increasingly recognizing the need to incorporate multidimensional perspectives to gain a comprehensive understanding of complex problems and make more informed decisions. By embracing diverse viewpoints, considering multiple dimensions, and leveraging advanced analytical techniques, organizations can capture a holistic view that leads to better outcomes. In this section, we will explore the importance of incorporating multidimensional perspectives and how it can enhance decision-making.

1. Understanding the Complexity of the Problem:

Many real-world problems are multifaceted, involving multiple variables, stakeholders, and interdependencies. Incorporating multidimensional perspectives allows decision-makers to grasp the complexity of the problem at hand. By considering diverse viewpoints and dimensions, organizations gain a deeper understanding of the various factors that influence the problem and its potential implications. This understanding enables decision-makers to make more comprehensive assessments and develop effective strategies that address the complexity of the problem.

2. Embracing Diversity and Inclusion:

Incorporating multidimensional perspectives promotes diversity and inclusion in decision-making processes. By engaging individuals with different backgrounds, expertise, and experiences, organizations tap into a wide range of perspectives. This diversity fosters creativity, innovation, and the exploration of alternative viewpoints. Decision-makers can leverage these diverse perspectives to challenge assumptions, overcome biases, and develop comprehensive solutions that consider a broad range of interests and needs.

3. Leveraging Advanced Analytical Techniques:

Advanced analytical techniques, such as data mining, machine learning, and network analysis, offer powerful tools to incorporate multidimensional perspectives. These techniques allow decision-makers to analyze and integrate data from multiple sources and dimensions. By leveraging these techniques, organizations can identify patterns, relationships, and trends that may not be apparent when considering a single perspective. This enhances decision-making by providing a more complete picture of the problem and facilitating the exploration of various scenarios and potential outcomes.

4. Stakeholder Engagement:

Incorporating multidimensional perspectives involves actively engaging stakeholders who hold diverse viewpoints and have a vested interest in the decision-making process. By involving stakeholders from different departments, functions, or external entities, organizations can gain insights into different aspects of the problem and consider a wide range of perspectives. This engagement promotes collaboration, builds trust, and increases the likelihood of implementing decisions successfully.

5. Considering Long-Term Implications:

A holistic view encompasses not only the immediate consequences of decisions but also their long-term implications. Incorporating multidimensional perspectives allows decision-makers to consider the broader impact of their choices on various stakeholders, the environment, and future generations. This long-term thinking enables organizations to make sustainable and responsible decisions that align with their values, mitigate potential risks, and create positive outcomes in the long run.

6. Balancing Trade-offs:

In complex decision-making scenarios, trade-offs are inevitable. Incorporating multidimensional perspectives enables decision-makers to identify and evaluate trade-offs across different dimensions. By considering various perspectives and priorities, organizations can strike a balance between conflicting objectives and make decisions that optimize outcomes across multiple dimensions. This balanced approach ensures a more inclusive and fair decision-making process.

Conclusion:

Incorporating multidimensional perspectives is essential for capturing a holistic view in decision-making. By embracing diverse viewpoints, leveraging advanced analytical techniques, engaging stakeholders, and considering long-term implications, organizations can make more informed and comprehensive assessments. This approach enables decision-makers to understand the complexity of problems, foster creativity and innovation, and develop strategies that optimize outcomes across multiple dimensions. As organizations prioritize multidimensional perspectives, they pave the way for more effective decision-making and create a positive impact in an increasingly interconnected and complex world.

Figure 5

Chapter 4: Scoring and Ranking Strategy Initiatives with Generative AI

Scoring and Ranking Strategy Initiatives with Generative AI: Unleashing the Power of Data-driven Decision-Making

In today's fast-paced and competitive business environment, organizations face the challenge of prioritizing and allocating resources to their strategy initiatives effectively. The abundance of data available poses both an opportunity and a challenge. However, by leveraging the power of generative AI, organizations can harness the potential of data-driven decision-making to score and rank their strategy initiatives more accurately and efficiently. In this section, we will explore how generative AI can revolutionize the process of scoring and ranking strategy initiatives, enabling organizations to make informed and impactful decisions.

Understanding Generative AI:

Generative AI utilizes advanced algorithms and models to generate new data based on existing patterns and relationships. It goes beyond traditional machine learning techniques by creating synthetic data that closely resembles the original dataset. Generative AI models, such as generative adversarial networks (GANs) and variational autoencoders (VAEs), can learn from vast amounts of data and generate new instances that adhere to the underlying data distribution. This capability allows organizations to leverage generative AI to enhance their decision-making processes.

Enhancing Scoring and Ranking with Generative AI

Generative AI can significantly enhance the scoring and ranking of strategy initiatives by incorporating additional insights and expanding the available dataset. Here's how generative AI can transform the process:

1. **Augmenting Data:** Generative AI enables organizations to augment their existing dataset by generating synthetic instances that closely resemble real-world data. By training generative models on historical data, organizations can create a larger and more diverse dataset, enhancing the accuracy and reliability of the scoring and ranking process. This augmented data allows decision-makers to evaluate initiatives against a broader range of scenarios, capturing a more comprehensive understanding of potential outcomes.

2. **Capturing Hidden Patterns:** Generative AI can uncover hidden patterns and relationships within the data that may not be apparent through traditional scoring methods. By generating synthetic data points that adhere to the underlying data distribution, generative models can reveal latent attributes, correlations, or dependencies that influence the success of strategy initiatives. This deeper understanding of the data empowers decision-makers to assess initiatives based on a holistic view, capturing factors that may have been previously overlooked.

3. **Simulating Potential Outcomes:** Generative AI enables organizations to simulate and evaluate a wide range of potential outcomes for their strategy initiatives. By generating synthetic data points and simulating different scenarios, decision-makers can assess the impact of various factors on the success of each initiative. This simulation-based approach provides a more comprehensive and accurate evaluation, allowing decision-makers to rank initiatives based on their potential performance under different conditions.

4. **Incorporating Risk Analysis:** Scoring and ranking strategy initiatives involves assessing and mitigating risks. Generative AI can enhance risk analysis by generating synthetic risk scenarios and simulating their impact on the outcomes of each initiative. This enables decision-makers to consider potential risks, evaluate mitigation strategies, and rank initiatives based on their resilience and ability to navigate uncertainties.

Generative AI empowers organizations to make more risk-aware decisions and prioritize initiatives that demonstrate robustness in the face of potential challenges.

5.

6. **Iterative Refinement:** Generative AI facilitates an iterative refinement process in scoring and ranking strategy initiatives. By generating synthetic data and simulating potential outcomes, decision-makers can evaluate the performance of different initiatives and refine their rankings based on new insights. This iterative approach enables organizations to adapt and adjust their rankings as they gather additional data, improving the accuracy and effectiveness of their decision-making.

Conclusion:

Generative AI offers unprecedented opportunities to enhance the scoring and ranking of strategy initiatives. By augmenting data, capturing hidden patterns, simulating potential outcomes, incorporating risk analysis, and enabling iterative refinement, organizations can make more informed and impactful decisions. Leveraging generative AI empowers decision-makers to take advantage of the vast amount of data available, uncover hidden insights, and evaluate strategy initiatives based on a comprehensive and data-driven approach. As organizations embrace generative AI, they will be better equipped to allocate resources effectively, prioritize initiatives accurately, and drive successful outcomes in today's dynamic and competitive business landscape.

Developing Scoring Frameworks with Generative AI Insights: Leveraging Data-driven Approaches for Enhanced Decision-Making

In the age of abundant data, organizations face the challenge of developing effective scoring frameworks to evaluate and prioritize their initiatives accurately. Traditional scoring methods often rely on predefined criteria and subjective assessments, limiting their ability to capture the full complexity of decision-making. However, with the advent of generative AI and its ability to uncover hidden patterns and relationships within data, organizations now have an unprecedented opportunity to develop scoring frameworks that leverage data-driven insights. In this section, we will explore how generative AI can enhance the development of scoring frameworks, enabling organizations to make more informed and objective decisions.

1. Augmenting Scoring Criteria:

Generative AI offers a powerful tool for augmenting scoring criteria within frameworks. By training generative models on historical data and generating synthetic data points, organizations can capture a broader range of factors that may impact the success of initiatives. These synthetic data points can reveal previously unidentified attributes, correlations, or dependencies that influence the scoring criteria. By incorporating generative AI insights, organizations can expand their scoring frameworks to encompass a more comprehensive set of criteria, resulting in a more accurate evaluation process.

2. Uncovering Hidden Patterns:

Generative AI is adept at uncovering hidden patterns and relationships within data that may go unnoticed by traditional scoring methods. By training generative models on vast amounts of data and analyzing the generated samples, organizations can gain valuable insights into underlying patterns and dynamics. These insights can be utilized to refine and enhance scoring frameworks. By incorporating generative AI insights, organizations can uncover new dimensions, interdependencies, or success factors that significantly impact the scoring process. This enables decision-makers to evaluate initiatives more holistically and capture nuanced aspects that were previously overlooked.

3. Simulating Scenarios and Outcomes:

Generative AI allows organizations to simulate and assess various scenarios and potential outcomes within their scoring frameworks. By generating synthetic data points and simulating different conditions, decision-makers can evaluate the performance and impact of each initiative under a range of circumstances. This simulation-based approach provides a more comprehensive and accurate evaluation, enabling organizations to make data-driven decisions and prioritize initiatives based on their potential outcomes. Generative AI insights empower decision-makers to consider the long-term implications and risks associated with different initiatives, leading to more informed and strategic choices.

4. Integrating Risk Analysis:

Risk assessment is a critical component of scoring frameworks, and generative AI can enhance this aspect. By generating synthetic risk scenarios and simulating their impact on the outcomes of initiatives, organizations can incorporate risk analysis directly into their scoring frameworks. Generative AI insights help decision-makers evaluate the resilience and robustness of initiatives in the face of potential risks. This integration of risk analysis within scoring frameworks ensures a more comprehensive evaluation, allowing decision-makers to prioritize initiatives that demonstrate a strong risk profile and align with the organization's risk appetite.

5. Iterative Refinement:

Generative AI facilitates an iterative refinement process in developing scoring frameworks. As organizations gather additional data and generate synthetic samples, decision-makers can continuously refine and update their scoring criteria. This iterative approach ensures that the scoring framework remains dynamic and adaptable to changing circumstances. Generative AI insights provide valuable feedback for fine-tuning the scoring criteria and improving the accuracy of the evaluation process over time.

Conclusion:

Developing scoring frameworks with generative AI insights empowers organizations to make more informed and objective decisions. By augmenting scoring criteria, uncovering hidden patterns, simulating scenarios and outcomes, integrating risk analysis, and enabling iterative refinement, organizations can create scoring frameworks that capture the complexity of decision-making. Leveraging generative AI enhances the objectivity, accuracy, and agility of scoring frameworks, leading to more effective evaluation and prioritization of initiatives. As organizations embrace generative AI in developing scoring frameworks, they gain a competitive advantage by leveraging data-driven insights and making more informed decisions in today's dynamic and data-rich business environment.

Combining Quantitative and Qualitative Factors in the Scoring Process: Achieving a Comprehensive Evaluation

In decision-making processes, both quantitative and qualitative factors play crucial roles in assessing and prioritizing options. While quantitative factors provide measurable and objective data, qualitative factors capture subjective aspects and nuanced insights. Recognizing the value of both approaches, organizations are increasingly combining quantitative and qualitative factors in the scoring process to achieve a more comprehensive evaluation. In this section, we will explore the benefits of combining these factors and how organizations can effectively integrate them to make informed decisions.

Understanding Quantitative and Qualitative Factors:

Quantitative factors involve measurable data that can be expressed numerically, such as financial metrics, performance indicators, or market data. These factors provide a quantitative foundation for analysis and decision-making. On the other hand, qualitative factors are subjective and involve non-measurable aspects, such as user experience, brand perception, or cultural fit. Qualitative factors capture insights, opinions, and subjective judgments that add depth and context to the decision-making process.

Benefits of Combining Quantitative and Qualitative Factors

1. **Comprehensive Evaluation:** By combining quantitative and qualitative factors, organizations can obtain a more comprehensive evaluation of options. Quantitative factors provide a quantitative baseline, while qualitative factors offer a more nuanced understanding of the strengths, weaknesses, and potential risks associated with each option. This holistic evaluation enables decision-makers to consider a wider range of factors, resulting in more robust and informed decisions.

2. **Balanced Decision-Making:** Integrating quantitative and qualitative factors helps balance objective and subjective considerations. While quantitative factors provide an objective assessment, qualitative factors address the subjective elements that can significantly impact outcomes. By considering both, decision-makers can account for diverse perspectives, opinions, and contextual factors, leading to well-rounded and balanced decision-making.

3. **Enhanced Risk Assessment:** Combining quantitative and qualitative factors enables organizations to assess risks more effectively. Quantitative data may highlight quantitative risks, such as financial instability or market volatility, while qualitative insights can identify qualitative risks like reputational damage or regulatory challenges. By integrating both types of factors, decision-makers gain a more comprehensive understanding of potential risks and can develop risk mitigation strategies accordingly.

4. **Improved Innovation and Creativity:** Qualitative factors foster creativity and innovation in decision-making. They encourage exploring unconventional ideas, user-centric perspectives, and unique value propositions. By integrating qualitative factors, organizations can consider factors such as market trends, customer preferences, or emerging technologies, which may not be captured by quantitative data alone. This integration promotes innovative thinking and helps organizations stay competitive in dynamic markets.

5. **Stakeholder Engagement and Alignment:** Combining quantitative and qualitative factors encourages stakeholder engagement and alignment. Quantitative factors provide a common language and framework for analysis, facilitating discussions and comparisons. Qualitative factors, on the other hand, allow stakeholders to express their insights, concerns, and preferences. This engagement helps build consensus, increase transparency, and foster a sense of ownership among stakeholders, leading to better decision acceptance and implementation.

Integrating Quantitative and Qualitative Factors

To effectively integrate quantitative and qualitative factors in the scoring process, organizations can follow these steps:

1. **Define Evaluation Criteria:** Clearly define the evaluation criteria and ensure a balance between quantitative and qualitative factors. Identify key performance indicators, financial metrics, and measurable benchmarks as quantitative criteria. Qualitative criteria can include user satisfaction, brand perception, cultural fit, or ethical considerations.

2. **Assign Weights and Scores:** Assign appropriate weights to each criterion based on its importance. Quantitative factors may have more objectively determined weights, while qualitative factors may require more subjective judgment. Develop scoring scales or rating systems that allow for consistent and comparative evaluation across different criteria.

3. **Gather Data and Insights:** Collect relevant quantitative data and conduct rigorous analysis to obtain measurable values for each criterion. For qualitative factors, gather insights through surveys, interviews, focus groups, or expert opinions. Capture qualitative data in a structured manner to ensure consistent evaluation.

4. **Evaluate and Compare Options:** Apply the defined evaluation criteria, weights, and scores to each option. Quantitative factors can be evaluated using mathematical calculations, while qualitative factors may involve subjective judgments. Consolidate the scores and rankings for a comprehensive evaluation and comparison of the options.

5. **Consider Trade-Offs and Decision Context:** Recognize that trade-offs may exist between quantitative and qualitative factors. Decision-makers should consider the specific context, strategic objectives, and the relative importance of each factor in the final decision. Transparently communicate the rationale behind the decision to stakeholders to maintain trust and foster alignment.

Conclusion:

Combining quantitative and qualitative factors in the scoring process allows organizations to achieve a more comprehensive and well-rounded evaluation of options. By integrating both types of factors, decision-makers gain a deeper understanding of risks, capture nuanced insights, foster innovation, and make more informed decisions. By effectively combining quantitative and qualitative factors, organizations can drive successful outcomes, adapt to changing environments, and remain competitive in today's dynamic business landscape.

Utilizing Generative Models for Sensitivity Analysis and Scenario-Based Scoring

Enhancing Decision-Making with Data-Driven Insights

In the realm of decision-making, organizations often face uncertainties, risks, and complex dynamics that can significantly impact the outcomes of their initiatives. To address these challenges, organizations are increasingly turning to generative models to perform sensitivity analysis and scenario-based scoring. Generative models, such as generative adversarial networks (GANs) and variational autoencoders (VAEs), offer a powerful approach to simulate diverse scenarios, assess the sensitivity of outcomes to different variables, and enhance decision-making with data-driven insights. In this section, we will explore how organizations can leverage generative models for sensitivity analysis and scenario-based scoring to make more informed and robust decisions.

Understanding Sensitivity Analysis and Scenario-Based Scoring:

Sensitivity analysis is a technique used to assess how sensitive the outcomes of a decision are to changes in key variables or parameters. It helps decision-makers understand the impact of uncertainties and assumptions on the results. Scenario-based scoring, on the other hand, involves evaluating different scenarios or alternative futures based on specific assumptions or conditions. This approach allows decision-makers to explore a range of possibilities and understand the potential outcomes under different circumstances.

Leveraging Generative Models for Sensitivity Analysis and Scenario-Based Scoring

Generative Models Offer Unique Capabilities That Can Enhance Sensitivity Analysis And Scenario-Based Scoring:

1. **Simulating Alternative Scenarios:** Generative models can generate synthetic data that closely resembles real-world scenarios. By training on historical data, generative models can generate new data points that represent alternative scenarios or potential futures. This allows decision-makers to simulate and evaluate the outcomes of different scenarios, providing valuable insights into the potential risks, opportunities, and trade-offs associated with each scenario.

2. **Exploring Variable Sensitivity:** Generative models can assess the sensitivity of outcomes to changes in key variables or parameters. By manipulating specific variables in the generative model, decision-makers can observe how the outcomes change in response to those variations. This sensitivity analysis enables organizations to identify the variables that have the most significant impact on the outcomes and make more informed decisions based on this understanding.

3. **Generating Synthetic Data for Scoring:** Generative models can generate synthetic data points that capture the characteristics and patterns of the underlying data distribution. This synthetic data can be used in the scoring process to evaluate different initiatives or options. By incorporating generative models' insights, decision-makers can gain a more comprehensive understanding of the potential outcomes, strengths, and weaknesses of each initiative. This data-driven scoring process enhances the objectivity and accuracy of decision-making.

4. **Optimizing Resource Allocation:** Generative models can assist in optimizing resource allocation by simulating and scoring different resource allocation strategies. By generating synthetic data points that reflect the relationships and dependencies among resources and initiatives, generative models enable decision-makers to evaluate the impact of different resource allocation decisions. This helps organizations allocate resources more effectively and prioritize initiatives based on their potential outcomes and resource requirements.

5. **Handling Uncertainties and Risk Assessment:** Generative models provide a framework for incorporating uncertainties and assessing risks in sensitivity analysis and scenario-based scoring. By generating synthetic data points that represent uncertain or risky scenarios, decision-makers can evaluate the potential risks and assess their impact on the outcomes. This data-driven approach enables organizations to make more risk-aware decisions and develop mitigation strategies based on a comprehensive analysis of different scenarios.

6. **Facilitating Iterative Refinement:** Generative models support an iterative refinement process in sensitivity analysis and scenario-based scoring. Decision-makers can update the generative model with new data and feedback, allowing for continuous improvement and adaptation of the scoring process. This iterative approach ensures that the sensitivity analysis and scenario-based scoring remain dynamic and responsive to changing circumstances and new insights.

Conclusion:

Generative models offer a powerful framework for performing sensitivity analysis and scenario-based scoring, enhancing decision-making processes. By simulating alternative scenarios, exploring variable sensitivity, generating synthetic data for scoring, optimizing resource allocation, handling uncertainties, and facilitating iterative refinement, organizations can make more informed and robust decisions. Leveraging generative models empowers decision-makers to assess the potential outcomes of different scenarios, evaluate sensitivities to key variables, and develop strategies that are resilient to uncertainties and risks. Ultimately, utilizing generative models for sensitivity analysis and scenario-based scoring enables organizations to navigate complex decision landscapes with greater confidence and achieve more successful outcomes.

Enhancing Accuracy and Reducing Biases in the Ranking of Strategy Initiatives: Leveraging Data-driven Approaches for Objective Decision-Making

Ranking strategy initiatives is a critical task for organizations aiming to allocate resources effectively and prioritize actions that align with their goals. However, this process can be challenging due to subjective biases, incomplete information, and inherent complexities. To address these challenges and enhance the accuracy of rankings, organizations are increasingly turning to data-driven approaches that leverage advanced analytical techniques. By embracing these approaches, organizations can reduce biases and make more objective, informed decisions. In this section, we will explore how organizations can enhance the accuracy and reduce biases in the ranking of strategy initiatives.

1. Data-driven Decision-Making:
Data-driven decision-making involves leveraging data and advanced analytical techniques to inform and guide the decision-making process. By collecting and analyzing relevant data, organizations can gain valuable insights into the performance, potential, and impact of strategy initiatives. This approach enables decision-makers to base their rankings on objective evidence rather than personal biases or assumptions.

2. Define Clear Evaluation Criteria:
To enhance accuracy, it is crucial to establish clear and well-defined evaluation criteria for ranking strategy initiatives. These criteria should align with the organization's objectives and be measurable, specific, and relevant. By establishing transparent criteria, decision-makers can ensure consistency and objectivity in the evaluation process. This clarity also helps mitigate biases that may arise from subjective interpretations or opinions.

3. Collect and Analyze Comprehensive Data:

To make informed decisions, organizations need access to comprehensive and reliable data. This includes both quantitative and qualitative data that captures various dimensions of the strategy initiatives. Quantitative data may involve financial metrics, performance indicators, market data, or customer feedback. Qualitative data, on the other hand, can provide insights into user experience, market trends, or competitive landscapes. By collecting and analyzing comprehensive data, decision-makers can obtain a holistic view of the initiatives and reduce biases caused by limited or incomplete information.

4. Incorporate Multiple Perspectives:

The inclusion of multiple perspectives in the ranking process can help reduce biases and ensure a more balanced evaluation. Involve stakeholders from different departments, functions, or levels of the organization to contribute their insights and perspectives. This diverse input helps avoid tunnel vision and promotes a more comprehensive assessment of the initiatives. By considering a range of perspectives, decision-makers can gain a broader understanding and minimize biases stemming from individual viewpoints.

5. Apply Statistical Analysis and Machine Learning:

Statistical analysis and machine learning techniques offer powerful tools to enhance accuracy and reduce biases in the ranking process. These techniques can identify patterns, correlations, and dependencies within the data, providing objective insights into the performance and potential of the initiatives. Regression analysis, clustering, decision trees, or predictive modeling can uncover hidden relationships and objectively assess the impact of different variables. By leveraging these techniques, organizations can make data-driven rankings that are less susceptible to biases and more accurate.

6. Conduct Sensitivity Analysis:

Sensitivity analysis helps assess the robustness of rankings by examining how changes in key variables or assumptions affect the results. By conducting sensitivity analysis, decision-makers can identify the most influential factors and understand the impact of uncertainties on the rankings. This analysis enables organizations to make more informed decisions and be prepared for potential variations or risks that may impact the initiatives' outcomes.

7. Continuously Evaluate and Update Rankings:

The accuracy of rankings can be improved by continuously evaluating and updating them as new data and insights become available. As strategy initiatives progress and new information emerges, decision-makers should review and revise the rankings accordingly. This adaptive approach ensures that the rankings remain relevant, accurate, and responsive to changing circumstances.

Conclusion:

Enhancing the accuracy and reducing biases in the ranking of strategy initiatives is crucial for organizations aiming to make objective, informed decisions. By embracing data-driven approaches, establishing clear evaluation criteria, collecting comprehensive data, incorporating multiple perspectives, applying statistical analysis and machine learning, conducting sensitivity analysis, and continuously evaluating rankings, organizations can improve the quality of their rankings and increase the likelihood of successful outcomes. By reducing biases and leveraging data-driven insights, organizations can navigate complex decision landscapes with greater confidence and drive the realization of their strategic objectives.

Figure 6

Chapter 5: Prioritization Strategies Enhanced by Generative AI

Harnessing the Power of Data-Driven Decision-Making

In the fast-paced and competitive business landscape, organizations often face the challenge of effectively prioritizing tasks, initiatives, and projects to optimize resource allocation and drive success. Traditional prioritization strategies can be time-consuming, subjective, and prone to biases. However, with the advent of generative AI, organizations now have a powerful tool at their disposal to enhance their prioritization strategies and make data-driven decisions.

In this section, we will explore how generative AI can revolutionize the process of prioritization, enabling organizations to allocate resources more efficiently and achieve their goals.

1. Generating Synthetic Data for Analysis:
Generative AI models, such as generative adversarial networks (GANs) and variational autoencoders (VAEs), have the capability to generate synthetic data that closely resembles real-

world datasets. By training these models on historical data, organizations can create additional data points to augment their existing dataset. This expanded dataset enables organizations to analyze a larger and more diverse pool of information, providing a more comprehensive understanding of the factors to consider during prioritization.

2. Capturing Complex Dependencies:

Prioritization often involves considering numerous factors and dependencies, which can be challenging to evaluate manually. Generative AI models excel at capturing complex dependencies within datasets. By leveraging generative AI insights, organizations can uncover hidden relationships, patterns, and dependencies that may not be apparent through traditional analysis methods. This enhanced understanding helps decision-makers make more informed choices when prioritizing initiatives, ensuring a holistic view of the impact and implications of each option.

3. Simulating Different Scenarios:

Generative AI models can simulate various scenarios based on synthesized data and real-world parameters. This simulation capability empowers organizations to evaluate the potential outcomes and impacts of different prioritization strategies. By generating synthetic data points and simulating scenarios, decision-makers can assess the effects of different resource allocations, project timelines, or market conditions on the prioritized initiatives. This allows organizations to make data-driven decisions, ensuring that resources are allocated strategically and initiatives are prioritized based on their potential impact and alignment with organizational goals.

4. Incorporating Uncertainty and Risk Analysis:

Prioritization decisions often involve inherent uncertainties and risks. Generative AI can aid in incorporating uncertainty and risk analysis into the prioritization process. By generating synthetic data that represents uncertainty factors and potential risk scenarios, decision-makers can assess the potential risks associated with each prioritized initiative. This data-driven approach allows organizations to consider risk mitigation strategies, evaluate the potential impact of uncertainties, and make more informed decisions while minimizing potential negative outcomes.

5. Iterative Refinement of Prioritization:

Generative AI enables organizations to continuously refine their prioritization strategies. As new data becomes available or as initiatives progress, decision-makers can update and refine the prioritization models based on generative AI insights. This iterative approach ensures that prioritization remains dynamic and adaptable to changing circumstances and new information. Decision-makers can refine their prioritization strategies based on the evolving business landscape, ensuring that resources are allocated optimally and align with current organizational objectives.

6. Mitigating Biases:

Generative AI can help mitigate biases that may influence prioritization decisions. By generating synthetic data points that represent a diverse range of scenarios and perspectives, generative AI models provide decision-makers with a more objective and unbiased view of the prioritization process. This enables organizations to make more fair and equitable decisions, ensuring that biases based on personal opinions or preferences are minimized.

Conclusion:

Generative AI offers significant potential for enhancing prioritization strategies in organizations. By generating synthetic data, capturing complex dependencies, simulating scenarios, incorporating uncertainty and risk analysis, facilitating iterative refinement, and mitigating biases, organizations can make data-driven prioritization decisions that optimize resource allocation and drive successful outcomes. Leveraging generative AI empowers decision-makers to make informed choices, ensuring that initiatives are prioritized strategically based on their potential impact, alignment with organizational goals, and consideration of uncertainties and risks. As organizations embrace generative AI to enhance their prioritization strategies, they gain a competitive advantage by harnessing the power of data-driven decision-making.

Leveraging Generative AI to Prioritize Strategy Initiatives Effectively: Enhancing Decision-Making in a Data-Driven World

In today's rapidly evolving business landscape, organizations are constantly challenged with the task of prioritizing strategy initiatives to achieve their goals efficiently. The abundance of data and the complexity of decision-making often make this process daunting and prone to biases. However, with the emergence of generative AI, organizations now have a powerful tool to leverage data-driven insights and enhance their ability to prioritize strategy initiatives effectively.

In this section, we will explore how generative AI can revolutionize the prioritization process, enabling organizations to allocate resources wisely and drive success.

1. Augmenting Data for Comprehensive Analysis:

Generative AI models, such as generative adversarial networks (GANs) and variational autoencoders (VAEs), have the ability to generate synthetic data that closely resembles real-world datasets. By training these models on historical data and using them to generate additional data points, organizations can augment their existing dataset. This expanded dataset enables decision-makers to analyze a more comprehensive range of information, leading to a more informed and accurate prioritization process.

2. Uncovering Hidden Patterns and Relationships:

Generative AI models excel at uncovering hidden patterns and relationships within datasets. By analyzing the synthesized and real-world data, organizations can gain insights into the complex dynamics and dependencies among various strategy initiatives. These insights enable decision-makers to understand the underlying factors that contribute to the success or failure of initiatives. By leveraging generative AI, organizations can identify patterns that may have been overlooked, ensuring a more comprehensive evaluation and effective prioritization.

3. Simulating Different Scenarios:
Generative AI allows organizations to simulate different scenarios and assess their potential outcomes. By generating synthetic data points and simulating various conditions, decision-makers can evaluate the impact of different resource allocations, market conditions, or external factors on the prioritized strategy initiatives. This simulation capability provides decision-makers with valuable insights, enabling them to make informed choices based on a comprehensive understanding of the potential consequences and trade-offs associated with each scenario.

4. Incorporating Data-Driven Metrics:
Generative AI enables organizations to incorporate data-driven metrics into the prioritization process. By analyzing the generated data points and utilizing advanced analytical techniques, organizations can develop metrics that objectively measure the potential value, feasibility, and alignment of strategy initiatives with organizational goals. These data-driven metrics help decision-makers quantify and compare the merits of different initiatives, facilitating a more informed and objective prioritization process.

5. Managing Risk and Uncertainty:
Prioritizing strategy initiatives involves inherent risks and uncertainties. Generative AI can assist organizations in managing these risks by generating synthetic data points that represent uncertain or risky scenarios. Decision-makers can evaluate the potential risks associated with each initiative, assess their impact, and develop risk mitigation strategies. By incorporating generative AI insights, organizations can make data-driven decisions, ensuring that risks are managed effectively and resources are allocated wisely to initiatives with the best risk-reward profiles.

6. Continuous Iteration and Improvement:
Generative AI facilitates a continuous iteration and improvement process in the prioritization of strategy initiatives. As new data becomes available or as initiatives progress, decision-makers can update the generative AI models and refine the prioritization criteria. This iterative approach ensures that the prioritization process remains dynamic, adaptable, and responsive to changing circumstances, leading to more accurate and effective decision-making.

Conclusion:
Generative AI offers organizations a powerful means to prioritize strategy initiatives effectively in a data-driven world. By augmenting data, uncovering hidden patterns, simulating scenarios, incorporating data-driven metrics, managing risks and uncertainties, and embracing continuous iteration, organizations can leverage the full potential of generative AI to make informed decisions. By utilizing generative AI, organizations can allocate resources wisely, maximize the chances of success, and drive positive outcomes in an increasingly complex and competitive business environment.

Incorporating Business Objectives, Constraints, and Risk Considerations: A Holistic Approach to Decision-Making

Effective decision-making requires organizations to consider various factors, including their business objectives, constraints, and risk considerations. By incorporating these elements into the decision-making process, organizations can ensure that their actions align with their strategic goals, account for limitations, and mitigate potential risks.

In this section, we will explore the importance of incorporating business objectives, constraints, and risk considerations, and how this holistic approach enhances decision-making.

1. Defining Business Objectives:

Business objectives serve as a compass, guiding organizations towards their desired outcomes. These objectives could include revenue growth, market expansion, cost reduction, customer satisfaction, or innovation. When making decisions, it is crucial to align choices with these objectives. By clearly defining business objectives, organizations can evaluate different options based on their potential to contribute to those objectives. This alignment ensures that decisions are purpose-driven and support the overall strategic direction of the organization.

2. Identifying Constraints:

Constraints refer to limitations or restrictions that organizations face when making decisions. These constraints can be related to resources, budget, time, regulations, or market conditions. It is essential to identify and consider these constraints during the decision-making process to ensure feasibility and practicality. By acknowledging constraints, organizations can assess the feasibility of various options, identify potential challenges, and make decisions that optimize available resources within the given limitations.

3. Assessing Risk Considerations:

Risk considerations play a crucial role in decision-making, as every decision carries a level of uncertainty and potential consequences. Organizations must evaluate and manage risks associated with different options. This includes identifying potential risks, assessing their impact on business objectives, and developing strategies to mitigate them. By incorporating risk considerations, organizations can make informed decisions that balance potential rewards with potential risks, ensuring that decisions are made with a comprehensive understanding of the associated uncertainties.

4. Utilizing Decision Frameworks and Tools:

To incorporate business objectives, constraints, and risk considerations effectively, organizations can employ decision frameworks and tools. These frameworks provide structured approaches to evaluate options, considering various factors in a systematic manner. For example, frameworks like the Analytical Hierarchy Process (AHP) or Decision Matrix enable organizations to assess different criteria, assign weights, and quantify the impact of options on business objectives. Decision support tools, such as decision trees or Monte Carlo simulations, can help quantify and evaluate risks associated with different choices. These frameworks and tools provide a structured approach to decision-making, ensuring that business objectives, constraints, and risk considerations are thoroughly considered.

5. Stakeholder Engagement and Collaboration:

Incorporating business objectives, constraints, and risk considerations requires collaboration and input from various stakeholders. Engaging stakeholders, including senior leadership, subject matter experts, and affected teams, ensures that different perspectives are considered. This collaborative approach enables organizations to identify additional constraints, assess risks more comprehensively, and align decisions with the broader goals and interests of the organization. By involving stakeholders, organizations can gain valuable insights, enhance decision quality, and foster a sense of ownership and commitment to the chosen course of action.

6. Regular Evaluation and Adaptation:

Incorporating business objectives, constraints, and risk considerations is an ongoing process. Organizations should regularly evaluate decisions in light of changing circumstances, evolving objectives, and emerging risks. This evaluation allows organizations to adapt their decisions and strategies accordingly, ensuring that they remain aligned with the business objectives and responsive to evolving constraints and risks. Regular evaluation and adaptation enable organizations to make agile decisions that reflect the dynamic nature of the business environment.

Conclusion:

Incorporating business objectives, constraints, and risk considerations into the decision-making process is crucial for organizations to make informed, strategic decisions. By aligning choices with business objectives, acknowledging constraints, assessing risks, utilizing decision frameworks and tools, engaging stakeholders, and regularly evaluating decisions, organizations can enhance their decision-making capabilities. This holistic approach ensures that decisions are purpose-driven, feasible, and well-informed, leading to improved outcomes and successful execution of business strategies.

Dynamic Prioritization Using Generative AI-Driven Optimization Technique: Maximizing Efficiency and Agility in Decision-Making

In today's fast-paced and ever-changing business landscape, organizations face the challenge of dynamically prioritizing their initiatives and tasks to stay competitive and adapt to evolving circumstances. Traditional prioritization methods often struggle to keep up with the dynamic nature of business, leading to suboptimal resource allocation and missed opportunities. However, by leveraging generative AI-driven optimization techniques, organizations can enhance their ability to dynamically prioritize and allocate resources effectively. In this section, we will explore how generative AI-driven optimization can revolutionize the prioritization process, enabling organizations to maximize efficiency, agility, and overall success.

1. Understanding Dynamic Prioritization:

Dynamic prioritization involves continuously reassessing and reprioritizing tasks and initiatives based on shifting circumstances, emerging opportunities, and changing business objectives. It requires organizations to be responsive and adaptive in allocating resources to align with their evolving goals. By adopting a dynamic approach, organizations can ensure that they are making the most efficient use of their resources and capitalizing on time-sensitive opportunities.

2. Leveraging Generative AI-Driven Optimization:

Generative AI-driven optimization techniques combine the power of generative AI models with optimization algorithms to identify the best possible allocation of resources in real-time. These techniques utilize historical and real-time data to generate synthetic scenarios, simulate resource allocation, and optimize prioritization decisions. By leveraging generative AI-driven optimization, organizations can harness the potential of vast amounts of data to make intelligent decisions and dynamically allocate resources for maximum impact.

3. Analyzing Complex Dependencies and Constraints:

Generative AI-driven optimization techniques excel at analyzing complex dependencies and constraints within the decision-making process. By considering a wide range of factors such as resource availability, dependencies among tasks, project timelines, and interdependencies between initiatives, organizations can identify optimal resource allocation strategies. This comprehensive analysis helps organizations make more informed decisions, considering the interplay of various factors and constraints.

4. Incorporating Real-Time Data and Feedback:

Dynamic prioritization requires organizations to incorporate real-time data and feedback into their decision-making process. Generative AI-driven optimization techniques enable organizations to integrate real-time data, such as market trends, customer feedback, or project progress, to adapt and reprioritize tasks accordingly. By leveraging real-time data and feedback, organizations can respond quickly to changing circumstances, make data-driven decisions, and allocate resources where they will have the most significant impact.

5. Continuous Learning and Adaptation:

Generative AI-driven optimization techniques facilitate continuous learning and adaptation in the prioritization process. As organizations gather new data and insights, generative AI models can be retrained and optimization algorithms can be refined to reflect the latest information. This iterative approach allows organizations to continuously improve their resource allocation strategies, adapt to evolving objectives and constraints, and stay ahead of the competition in a dynamic business environment.

6. Balancing Short-Term and Long-Term Priorities:

Effective dynamic prioritization requires striking a balance between short-term and long-term priorities. While short-term goals may demand immediate attention, long-term objectives and strategic initiatives should not be neglected. Generative AI-driven optimization techniques help organizations find the optimal balance by considering both short-term and long-term priorities. By incorporating both perspectives, organizations can allocate resources strategically to achieve immediate results while advancing their long-term goals.

7. Facilitating Collaboration and Transparency:

Generative AI-driven optimization techniques facilitate collaboration and transparency in the prioritization process. These techniques provide a clear framework for decision-making and enable stakeholders to collaborate, provide inputs, and contribute to the prioritization process. This collaboration fosters transparency, aligns stakeholders, and ensures that decisions are made with a comprehensive understanding of the factors and considerations involved.

Conclusion:

Dynamic prioritization is crucial for organizations seeking to maximize efficiency and agility in their decision-making processes. By leveraging generative AI-driven optimization techniques, organizations can adapt to changing circumstances, allocate resources effectively, and make informed decisions in real-time. This approach enables organizations to optimize resource utilization, capitalize on opportunities, and drive overall success in today's dynamic business landscape. By embracing generative AI-driven optimization, organizations can unlock the full potential of dynamic prioritization and enhance their competitive advantage.

Uncovering Novel and Unconventional Prioritization Strategies with Generative AI: Unlocking New Avenues for Decision-Making

In the realm of decision-making, organizations are constantly seeking innovative approaches to prioritize their initiatives effectively. Traditional prioritization strategies may not always capture the full spectrum of possibilities and opportunities. However, with the emergence of generative AI, organizations now have a powerful tool to uncover novel and unconventional prioritization strategies. By leveraging generative AI, organizations can break free from conventional thinking and explore new avenues for decision-making. In this section, we will explore how generative AI can revolutionize the prioritization process, enabling organizations to uncover unique strategies that can drive success.

1. Divergent Thinking and Idea Generation:

Generative AI models, such as generative adversarial networks (GANs) and variational autoencoders (VAEs), have the capability to generate novel and diverse ideas by simulating alternative scenarios. By training these models on vast datasets and leveraging their generative capabilities, organizations can explore a wide range of possibilities that go beyond traditional thinking. This divergent thinking approach helps organizations break free from conventional constraints and uncover unconventional ideas and strategies for prioritization.

2. Simulating Unexplored Scenarios:

Generative AI allows organizations to simulate unexplored scenarios and evaluate their potential outcomes. By generating synthetic data points that represent alternative realities or emerging trends, decision-makers can assess the potential impact and feasibility of prioritizing initiatives in these scenarios. This simulation capability enables organizations to uncover potential opportunities and risks that may have been overlooked using traditional approaches. By exploring unexplored scenarios, organizations can identify unique strategies that have the potential to drive innovation and competitive advantage.

3. Identifying Emerging Patterns and Trends:

Generative AI models have the ability to identify emerging patterns and trends within datasets. By analyzing large volumes of data and leveraging advanced machine learning techniques, these models can detect subtle signals and identify emerging opportunities or market shifts. This insight helps decision-makers prioritize initiatives that align with these emerging trends, allowing organizations to stay ahead of the curve and capitalize on new opportunities. By leveraging generative AI, organizations can identify unconventional strategies that are driven by emerging patterns and trends, providing a competitive edge in rapidly evolving industries.

4. Collaborative Exploration and Idea Generation:

Generative AI can facilitate collaborative exploration and idea generation among teams and stakeholders. By leveraging generative AI models as a creative tool, organizations can engage employees, subject matter experts, and stakeholders in the process of generating and refining ideas for prioritization. This collaborative approach fosters diverse perspectives, encourages cross-functional collaboration, and unlocks collective intelligence. By incorporating the insights and ideas generated through generative AI, organizations can uncover unique and unconventional strategies that tap into the collective wisdom of their teams.

5. Combining Quantitative and Qualitative Factors:

Generative AI allows organizations to combine quantitative and qualitative factors in the prioritization process. While traditional approaches often focus on measurable metrics and objective data, generative AI can incorporate subjective factors and unstructured data, such as user feedback, market sentiment, or expert opinions. By integrating both quantitative and qualitative factors, organizations can make more holistic and comprehensive decisions that consider a broader range of perspectives. This approach enables the exploration of unconventional strategies that may be overlooked when relying solely on traditional, quantitative methods.

6. Iterative Refinement and Adaptation:

Generative AI facilitates iterative refinement and adaptation in the prioritization process. As organizations gather feedback and new data, generative AI models can be trained and refined to incorporate the latest insights. This iterative approach allows organizations to continuously improve their prioritization strategies, adapt to changing circumstances, and uncover novel strategies that evolve with the dynamic business landscape.

Conclusion:

Generative AI offers organizations a powerful tool to uncover novel and unconventional prioritization strategies. By leveraging generative AI models to stimulate divergent thinking, simulate unexplored scenarios, identify emerging patterns, foster collaborative exploration, combine quantitative and qualitative factors, and enable iterative refinement, organizations can unlock new avenues for decision-making. These unconventional strategies can lead to breakthrough innovations, competitive advantage, and the realization of strategic objectives. By embracing generative AI, organizations can tap into the full potential of their data and expertise, driving success in an ever-evolving business environment.

"There are some ethical roadblocks,
so do we use our Greedy Positioning System
or our moral compass to find a detour??"

Figure 7

Chapter 6: Ethical Considerations in Generative AI-Driven Decision Intelligence

Navigating the Intersection of Technology and Ethics:

Generative AI has revolutionized decision intelligence, empowering organizations to make data-driven and informed decisions. However, as organizations leverage generative AI to enhance decision-making processes, it is crucial to acknowledge and address the ethical considerations that arise in this context. Ethical considerations play a pivotal role in ensuring that generative AI-driven decision intelligence is deployed responsibly, transparently, and in alignment with societal values. In this section, we will explore the ethical considerations associated with generative AI-driven decision intelligence and discuss the importance of navigating the intersection of technology and ethics.

1. Data Privacy and Security:

Generative AI relies on large datasets to generate insights and simulate scenarios. Organizations must prioritize data privacy and security to protect sensitive information. It is crucial to obtain appropriate consent, anonymize data, and implement robust security measures to prevent unauthorized access or misuse of data. Organizations must be transparent with stakeholders about data collection, usage, and storage practices to foster trust and respect privacy rights.

2. Algorithmic Bias and Fairness:

Generative AI models learn from historical data, which may reflect biases present in society. Organizations must be vigilant in identifying and mitigating algorithmic bias to ensure fairness and equity in decision-making processes. This involves assessing training data for potential biases, testing models for fairness, and implementing strategies to reduce bias impact. Regular monitoring and evaluation are necessary to address any biases that may emerge during the generative AI-driven decision intelligence process.

3. Explainability and Transparency:

Generative AI models can be complex and opaque, making it challenging to understand the reasoning behind their outputs. Decision intelligence should strive for explainability and transparency to build trust and accountability. Organizations must ensure that decision-making processes are transparently communicated, providing insights into how generative AI models operate, the factors influencing decisions, and the limitations of the technology. Explainability helps stakeholders understand and challenge the outcomes, fostering ethical decision-making practices.

4. Human Oversight and Responsibility:

While generative AI can augment decision intelligence, human oversight and responsibility are essential. Organizations should ensure that humans are involved in the decision-making process and exercise critical judgment. Humans play a crucial role in defining the ethical framework, setting objectives, validating outputs, and considering the broader social implications of decisions. It is important to strike a balance between the capabilities of generative AI and human judgment to ensure ethical and responsible decision intelligence.

5. Accountability and Auditing:

Accountability is essential in generative AI-driven decision intelligence. Organizations should establish clear lines of responsibility and accountability for decisions made using generative AI models. Regular audits should be conducted to evaluate the fairness, accuracy, and impact of the decisions. This ensures that decisions align with ethical standards, legal requirements, and organizational values. Organizations should be open to external scrutiny, encouraging independent audits and assessments of their generative AI-driven decision intelligence practices.

6. Continuous Ethical Evaluation:

Ethical considerations in generative AI-driven decision intelligence evolve over time. Organizations must engage in continuous ethical evaluation and reflection to adapt to changing societal norms, legal frameworks, and stakeholder expectations. This includes staying abreast of emerging ethical guidelines, participating in industry discussions, and seeking external expertise to ensure that decision intelligence practices align with the evolving ethical landscape.

Conclusion:

Generative AI-driven decision intelligence holds great potential for organizations, but it also raises important ethical considerations. By prioritizing data privacy and security, addressing algorithmic bias, fostering explainability and transparency, maintaining human oversight, ensuring accountability, and engaging in continuous ethical evaluation, organizations can navigate the intersection of technology and ethics responsibly. By integrating ethical considerations into generative AI-driven decision intelligence, organizations can harness the benefits of this powerful technology while upholding ethical values and societal expectations.

Addressing Ethical Challenges in the Utilization of Generative AI for Decision Intelligence: A Responsible Approach

Generative AI has emerged as a powerful tool for decision intelligence, offering organizations the ability to analyze data, simulate scenarios, and make informed choices. However, as with any transformative technology, the utilization of generative AI for decision intelligence raises ethical challenges that must be addressed to ensure responsible and ethical deployment. In this section, we will explore some of the key ethical challenges associated with generative AI and discuss strategies for addressing them in the pursuit of responsible decision intelligence.

1. Data Privacy and Consent:

One of the primary ethical challenges in utilizing generative AI is safeguarding data privacy and obtaining informed consent. Generative AI models rely on vast amounts of data, often collected from individuals. Organizations must prioritize data privacy, ensuring that data is handled securely and anonymized to protect individuals' identities. Additionally, obtaining clear and informed consent from data subjects is crucial to ensure transparency and respect for their rights.

2. Algorithmic Bias and Fairness:

Generative AI models are trained on historical data, which can introduce biases into the decision-making process. It is essential to address algorithmic bias and ensure fairness in decision intelligence. Organizations should assess training data for biases, actively mitigate them, and regularly evaluate model outputs to identify and rectify any potential biases that arise. By striving for fairness and equity, organizations can ensure that generative AI-driven decision intelligence benefits all stakeholders.

3. Transparency and Explainability:

Generative AI models can be complex and opaque, making it difficult to understand the rationale behind their decisions. Ensuring transparency and explainability is essential to address ethical concerns. Organizations should aim to make their decision-making processes transparent, providing explanations of how generative AI models arrive at their outputs. This transparency fosters trust, enables stakeholders to understand the reasoning behind decisions, and promotes accountability.

4. Human Oversight and Responsibility:

While generative AI can augment decision intelligence, human oversight and responsibility are critical. Humans must retain control and exercise judgment throughout the decision-making process. Organizations should establish clear lines of responsibility and accountability, ensuring that humans play an active role in defining the objectives, validating outputs, and considering the ethical implications of decisions. Human oversight helps ensure that generative AI is used responsibly and aligns with ethical standards.

5. Auditing and Accountability:

To address ethical challenges, organizations should establish auditing mechanisms and ensure accountability for generative AI-driven decision intelligence. Regular audits should be conducted to evaluate the fairness, accuracy, and impact of decisions. This enables organizations to identify and rectify any ethical concerns that arise, ensuring that decisions align with legal requirements, ethical standards, and organizational values. Accountability mechanisms promote responsible utilization of generative AI in decision intelligence.

6. Continuous Ethical Assessment and Adaptation:

Ethical considerations in generative AI-driven decision intelligence evolve over time. Organizations should engage in continuous ethical assessment and adaptation, keeping abreast of emerging ethical guidelines, industry best practices, and societal expectations. By seeking external expertise, participating in ethical discussions, and regularly reflecting on the ethical implications of their practices, organizations can proactively address ethical challenges and ensure responsible decision intelligence.

Conclusion:

As generative AI becomes increasingly integrated into decision intelligence processes, organizations must address ethical challenges to ensure responsible and ethical deployment. By prioritizing data privacy and consent, addressing algorithmic bias, promoting transparency and explainability, maintaining human oversight, establishing auditing mechanisms, and engaging in continuous ethical assessment, organizations can navigate the ethical landscape associated with generative AI. By adopting a responsible approach, organizations can harness the benefits of generative AI while upholding ethical values and safeguarding the interests of stakeholders.

Ensuring Fairness, Transparency, and Accountability in Assessments and Scoring: Building Trust in Decision-Making Processes

In an increasingly data-driven world, assessments and scoring systems play a crucial role in decision-making across various sectors, including education, employment, healthcare, and finance. Ensuring these systems are fair, transparent, and accountable is essential for building trust and fostering equity. This article explores the importance of these principles and strategies for their implementation.

The Importance of Fairness in Assessments

Fairness in assessments ensures that every individual is evaluated based on objective criteria, free from biases or discrimination. This is critical for maintaining the integrity of any assessment system and for ensuring that outcomes are just and equitable. Unfair assessments can lead to significant adverse impacts, including lost opportunities and perpetuated inequalities.

Strategies for Ensuring Fairness

1. **Standardization:** Implementing standardized procedures and criteria helps minimize subjective biases. This involves using consistent methods and metrics for all individuals being assessed.
2. **Bias Mitigation:** Training for assessors on unconscious biases and implementing tools to detect and correct these biases can promote fairness.
3. **Inclusive Design:** Designing assessments that consider the diverse backgrounds and needs of individuals ensures that no group is systematically disadvantaged.

The Role of Transparency in Building Trust

Transparency in assessment processes involves openly sharing how decisions are made and on what basis scores are assigned. When individuals understand the criteria and processes, they are more likely to trust the outcomes.

Strategies for Ensuring Transparency

1. **Clear Communication:** Providing detailed information about assessment criteria, scoring methods, and decision-making processes in accessible language.
2. **Open Acces:** Allowing individuals to review their assessments and understand their scores can help demystify the process.
3. **Regular Audits:** Conducting and publishing regular audits of assessment systems to verify that they are functioning as intended.

Accountability in Assessment Systems

Accountability ensures that those responsible for assessments are answerable for their actions and decisions. This principle is essential for maintaining public trust and for continuous improvement of the assessment systems.

Strategies for Ensuring Accountability

1. **Independent Oversight:** Establishing independent bodies to oversee the assessment processes and handle grievances can enhance accountability.
2. **Feedback Mechanisms:** Implementing robust feedback systems where individuals can report issues or unfair practices.
3. **Continuous Improvement:** Regularly reviewing and updating assessment systems

based on feedback and changing needs to ensure they remain relevant and fair.

4.

Case Studies and Best Practices

Education Sector

In education, standardized testing is often scrutinized for fairness and transparency. For instance, the College Board, which administers the SAT, has made efforts to redesign the test to be more reflective of high school curricula and to reduce biases. They also provide detailed score reports and have an appeals process for contested scores.

Employment

In recruitment, companies are increasingly using AI and machine learning algorithms for candidate assessments. To ensure fairness, companies like Unilever have adopted structured interviews and standardized assessment tools. They also ensure transparency by providing candidates with feedback and detailed explanations of their assessment results.

Healthcare

In healthcare, scoring systems like the APACHE (Acute Physiology and Chronic Health Evaluation) are used to predict patient outcomes. Ensuring fairness involves regularly validating these models across different patient demographics. Transparency and accountability are maintained by publishing the methodologies and results of these validations in peer-reviewed journals.

The Path Forward

Ensuring fairness, transparency, and accountability in assessments and scoring systems is an ongoing process that requires commitment from all stakeholders. By adopting standardized practices, promoting inclusivity, and maintaining open channels of communication, organizations can build trust and ensure that their decision-making processes are perceived as just and equitable.

Conclusion

As we continue to rely on assessments and scoring systems for critical decisions, it is imperative to embed fairness, transparency, and accountability into these processes. Doing so not only builds trust but also ensures that we move towards a more equitable and just society. Through continuous effort and vigilance, we can create assessment systems that truly reflect the principles of fairness and equity.

Mitigating Biases and Discriminatory Outcomes: Promoting Fairness and Equity in Decision-Making Processes

In decision-making processes, biases and discriminatory outcomes can undermine fairness, equity, and trust. Whether it's in hiring practices, loan approvals, educational evaluations, or any other area where decisions impact individuals' lives, it is essential to actively mitigate biases and ensure equitable outcomes. By recognizing and addressing biases, organizations can foster inclusive and unbiased decision-making processes. In this section, we will explore strategies for mitigating biases and promoting fairness and equity in decision-making.

1. Recognizing Implicit and Explicit Biases:

The first step in mitigating biases is to recognize and acknowledge their presence. Biases can be explicit or implicit, and they may arise from conscious or unconscious attitudes and beliefs. Organizations should invest in awareness programs and training to help decision-makers understand their own biases and recognize how they can influence their judgment. Recognizing biases is the foundation for addressing and mitigating them effectively.

2. Diverse and Inclusive Decision-Making Panels:

Diverse perspectives and experiences are crucial in minimizing biases. Organizations should ensure that decision-making panels and committees reflect diverse backgrounds, including different genders, ethnicities, cultures, and abilities. By incorporating a range of perspectives, biases can be challenged and corrected, leading to more equitable decisions. Additionally, involving individuals who have expertise in diversity and inclusion can contribute to more inclusive decision-making processes.

3. Structured Decision-Making Processes:

Structuring decision-making processes can help mitigate biases by ensuring consistency and transparency. Clear criteria, guidelines, and scoring rubrics should be established to evaluate individuals objectively. Structured processes reduce the influence of subjective biases and personal preferences, leading to fairer and more equitable outcomes. Decision-makers should be trained on how to apply these criteria consistently, minimizing the potential for bias to seep into the decision-making process.

4. Continuous Evaluation and Improvement:

Mitigating biases requires ongoing evaluation and improvement. Organizations should regularly review their decision-making processes, monitor outcomes, and assess any potential biases. This includes analyzing data for disparate impacts on different groups, examining decision patterns, and conducting statistical analyses to identify biases. By continuously evaluating and refining decision-making processes, organizations can address biases proactively and foster more equitable outcomes.

5. Robust Data Collection and Analysis:

Data collection and analysis are critical for identifying biases and discriminatory outcomes. Organizations should collect and analyze data on decision outcomes, demographics, and other relevant variables to identify any disparities among different groups. This analysis can uncover biases that may not be apparent on the surface and enable organizations to take corrective actions.

By incorporating data-driven insights, organizations can make informed decisions and work towards equitable outcomes.

6. Implementing Bias-Mitigating Technologies:

Technological tools, such as AI algorithms and machine learning models, can help identify and mitigate biases in decision-making processes. These tools can analyze data patterns, detect potential biases, and provide recommendations to decision-makers to reduce bias. However, it is important to ensure that these technologies are developed and trained with diverse and representative datasets to avoid perpetuating existing biases. Human oversight is also crucial to ensure that technology is used responsibly and ethically.

7. Regular Training and Education:

Ongoing training and education are vital to building awareness and skills to mitigate biases effectively. Decision-makers should receive training on unconscious bias, diversity and inclusion, cultural competence, and ethical decision-making. These programs should emphasize the importance of fairness, equity, and inclusivity in decision-making processes and provide practical strategies for recognizing and addressing biases. Regular training helps decision-makers stay informed and proactive in their efforts to mitigate biases.

Conclusion:

Mitigating biases and discriminatory outcomes is a critical endeavor in promoting fairness and equity in decision-making processes. By recognizing biases, promoting diverse and inclusive decision-making panels, structuring processes, evaluating outcomes, utilizing data-driven insights, implementing bias-mitigating technologies, and providing regular training, organizations can take meaningful steps toward creating more equitable and unbiased decision-making environments. By prioritizing fairness and equity, organizations can contribute to a more inclusive society and foster trust and confidence in their decision-making processes.

Establishing Ethical Guidelines for Responsible Generative AI-Driven Decision-Making: Upholding Integrity and Accountability

As generative AI continues to evolve and shape decision-making processes, it is crucial to establish ethical guidelines that govern its responsible use. Generative AI-driven decision-making holds great potential for organizations, but it also raises ethical considerations that must be addressed to ensure integrity, fairness, and accountability. By setting clear ethical guidelines, organizations can navigate the complexities of generative AI-driven decision-making and promote responsible practices. In this section, we will explore the importance of establishing ethical guidelines and discuss key principles for responsible generative AI-driven decision-making.

1. Transparency and Explainability:

Transparency and explainability are fundamental principles in responsible generative AI-driven decision-making. Organizations should ensure that the decision-making process, including the use of generative AI models, is transparently communicated to stakeholders. This involves providing insights into the model's operations, the data used, the decision criteria, and the limitations of the technology. Explainability enables stakeholders to understand the decision-making process and fosters trust and accountability.

2. Data Privacy and Consent:

Responsible generative AI-driven decision-making requires a strong commitment to data privacy and obtaining informed consent. Organizations should prioritize the protection of individuals' data, ensuring that it is collected, stored, and used in compliance with applicable privacy laws and regulations. Obtaining clear and informed consent from individuals whose data is utilized is essential to respect their rights and maintain trust.

3. Fairness and Bias Mitigation:

Fairness is a critical ethical consideration in generative AI-driven decision-making. Organizations should actively work to identify and mitigate biases that may be present in the data or models. This involves conducting regular audits, monitoring for disparate impacts on different groups, and implementing strategies to reduce bias. Fairness should be a guiding principle throughout the decision-making process to ensure equitable outcomes for all individuals.

4. Accountability and Oversight:

Establishing accountability and oversight mechanisms is essential for responsible generative AI-driven decision-making. Organizations should define clear lines of responsibility and ensure that decision-makers are accountable for the outcomes of their decisions. Regular audits, evaluations, and quality assurance processes help ensure that decisions align with ethical standards and organizational values. External audits or third-party assessments can provide additional accountability and transparency.

5. Human Oversight and Intervention:

Human oversight and intervention play a crucial role in responsible decision-making processes involving generative AI. While generative AI models can provide valuable insights, human judgment is essential in interpreting and contextualizing the outputs. Organizations should ensure that human decision-makers are actively involved throughout the process, exercising critical thinking, and considering ethical implications. Human intervention allows for context-specific considerations and ethical reasoning that may not be captured by the generative AI models alone.

6. Continuous Evaluation and Improvement:

Responsible generative AI-driven decision-making requires continuous evaluation and improvement. Organizations should regularly assess the impact of generative AI models on decision outcomes, monitor for biases or unintended consequences, and adapt their practices accordingly. By embracing a culture of continuous learning and improvement, organizations can enhance the ethical integrity of their decision-making processes over time.

7. Collaboration and Stakeholder Engagement:

Incorporating diverse perspectives and engaging stakeholders is crucial for responsible generative AI-driven decision-making. Organizations should actively seek input from individuals who are affected by the decisions, as well as experts in the field. Engaging stakeholders promotes inclusivity, provides valuable insights, and helps identify potential ethical concerns. Collaboration ensures that decisions reflect a broader range of perspectives and contribute to collective decision-making processes.

Conclusion:

Establishing ethical guidelines is vital to ensure responsible generative AI-driven decision-making. By prioritizing transparency, data privacy, fairness, accountability, human oversight, continuous evaluation, and stakeholder engagement, organizations can navigate the ethical challenges associated with generative AI. Responsible generative AI-driven decision-making upholds integrity, promotes equity, and safeguards against unintended consequences. By adhering to ethical guidelines, organizations can harness the power of generative AI while maintaining trust, accountability, and societal benefit.

Figure 8

Chapter 7: Overcoming Implementation Challenges with Implementing Generative AI

Strategies for Success:

Generative AI holds tremendous potential for transforming industries and revolutionizing decision-making processes. However, organizations often encounter various implementation challenges when integrating generative AI into their workflows. From technical complexities to cultural shifts, these challenges can impede progress and hinder the realization of the full benefits of generative AI. In this section, we will explore common implementation challenges with generative AI and discuss strategies to overcome them, enabling organizations to successfully integrate and leverage this powerful technology.

1. Data Availability and Quality:
One of the key challenges in implementing generative AI is accessing relevant and high-quality data. Generative AI models require extensive and diverse datasets to learn and generate meaningful outputs. Organizations must identify and gather the necessary data, ensuring it is comprehensive,

representative, and free from biases. Data collection, cleansing, and validation processes should be established to ensure the accuracy and reliability of the data used in generative AI models.

Strategy: Collaboration and Partnerships - Organizations can collaborate with industry partners, academic institutions, or third-party data providers to access a broader range of data sources. Partnerships can help overcome data scarcity challenges and ensure a more comprehensive and diverse dataset for training generative AI models.

2. Computational Power and Infrastructure:

Generative AI models often require significant computational power and infrastructure to train and deploy effectively. The complexity of these models can strain existing IT systems and infrastructure, resulting in performance limitations and increased costs. Organizations must assess their computational capabilities and invest in the necessary hardware, software, and cloud infrastructure to support the implementation of generative AI.

Strategy: Scalable Infrastructure and Cloud Computing - Leveraging cloud-based infrastructure and platforms can provide the necessary scalability and computing power to handle the demands of generative AI. Cloud solutions offer flexibility, cost-effectiveness, and the ability to scale resources as needed, minimizing the infrastructure challenges associated with implementing generative AI.

3. Model Interpretability and Explainability:

Generative AI models, such as deep neural networks, are often complex and difficult to interpret or explain. This lack of interpretability can raise concerns, especially in regulated industries or applications where transparency is crucial. Organizations must address the challenge of interpreting and explaining the decisions or outputs generated by generative AI models to build trust and ensure regulatory compliance.

Strategy: Hybrid Approaches and Explainability Techniques - Organizations can adopt hybrid approaches that combine generative AI models with interpretable models. This allows decision-makers to understand the rationale behind the outputs generated by generative AI. Additionally, research and development efforts should focus on developing explainability techniques specific to generative AI, enabling better understanding and transparency.

4. Change Management and Cultural Shifts:

Implementing generative AI involves a cultural shift within organizations. Resistance to change, lack of awareness, and unfamiliarity with the technology can hinder successful implementation. Employees may require new skill sets and training to effectively work with generative AI systems, and management support is crucial to drive adoption and overcome resistance.

Strategy: Training and Education Programs - Organizations should invest in training and education programs to build awareness and understanding of generative AI among employees. Training programs should include technical skills development as well as awareness of the benefits and limitations of generative AI. Engaging employees through workshops, knowledge-sharing sessions, and hands-on experiences can foster a culture of acceptance and facilitate a smoother implementation process.

5. Ethical and Legal Considerations:
Generative AI raises ethical and legal considerations, such as data privacy, bias mitigation, and regulatory compliance. Organizations must navigate these challenges to ensure responsible and ethical use of generative AI, aligning with legal requirements and societal expectations.

Strategy: Ethical Frameworks and Governance Policies - Organizations should establish clear ethical frameworks and governance policies that guide the use of generative AI. This includes implementing data privacy measures, conducting regular audits to identify and mitigate biases, and complying with relevant regulations and industry standards. Collaborating with legal and ethics experts can help organizations develop robust frameworks and policies that address the specific challenges associated with generative AI.

Conclusion:
Overcoming implementation challenges is crucial for organizations to successfully integrate generative AI into their workflows and unlock its full potential. By addressing data availability, computational power, interpretability, change management, and ethical considerations, organizations can pave the way for a successful implementation of generative AI. With careful planning, strategic investments, and a focus on ethical and responsible practices, organizations can harness the transformative power of generative AI and drive innovation in their respective industries.

Data Acquisition and Preparation for Generative AI-Driven Decision Intelligence: Building a Strong Foundation

Generative AI-driven decision intelligence relies heavily on data - the fuel that powers the models and enables insightful decision-making. However, data acquisition and preparation present significant challenges that organizations must overcome to ensure the accuracy, reliability, and effectiveness of generative AI models. In this section, we will explore the importance of data acquisition and preparation for generative AI-driven decision intelligence and discuss strategies for building a strong foundation.

1. Defining Data Requirements:
The first step in data acquisition is to clearly define the data requirements for generative AI-driven decision intelligence. This involves identifying the specific types of data needed, such as structured

or unstructured data, text, images, or time series data. It also entails determining the volume, variety, and velocity of data necessary to train and validate the generative AI models effectively. By clearly defining data requirements, organizations can focus their efforts on acquiring the most relevant and valuable data for decision intelligence.

2. Data Sources and Partnerships:

Acquiring high-quality data can be a significant challenge. Organizations should identify and establish partnerships with relevant data sources and providers to access diverse and comprehensive datasets. This may involve collaborating with external organizations, academic institutions, or data vendors who specialize in specific domains or industries. By leveraging partnerships, organizations can expand their data sources and gain access to valuable datasets that enhance the effectiveness of generative AI-driven decision intelligence.

3. Data Collection and Cleansing:

Once data sources are identified, organizations must collect and cleanse the data to ensure its accuracy and reliability. Data collection processes should be designed to capture relevant information, leveraging appropriate tools and technologies for data extraction, such as web scraping, APIs, or surveys. Data cleansing involves removing duplicate entries, handling missing values, and addressing outliers or anomalies. Thorough data cleansing is crucial to ensure the quality and integrity of the data used for training generative AI models.

4. Data Annotation and Labeling:

Generative AI models often require annotated or labeled data to learn patterns and generate meaningful outputs. Data annotation involves labeling data instances with relevant attributes or categories. Organizations may need to develop annotation guidelines, establish annotation teams, or leverage crowd-sourcing platforms to annotate large datasets efficiently. Accurate and consistent annotation is essential for training generative AI models and improving decision intelligence outcomes.

5. Data Privacy and Compliance:

Data acquisition and preparation must adhere to data privacy regulations and ethical considerations. Organizations must ensure that data acquisition processes comply with relevant laws and regulations, such as the General Data Protection Regulation (GDPR) or Health Insurance Portability and Accountability Act (HIPAA), depending on the data being collected. Implementing proper data privacy measures, obtaining consent, anonymizing sensitive information, and establishing data governance practices are crucial for maintaining compliance and protecting individuals' privacy.

6. Data Augmentation and Enrichment:

To enhance the diversity and richness of the data used for generative AI-driven decision intelligence, organizations can leverage data augmentation and enrichment techniques. Data augmentation involves artificially expanding the dataset by creating variations of existing data through techniques like rotation, scaling, or adding noise. Data enrichment involves enhancing the dataset by integrating external data sources, such as demographic data, market trends, or expert opinions. These techniques can help generate more robust and representative models, improving decision intelligence capabilities.

7. Data Management and Accessibility:
Efficient data management is vital for generative AI-driven decision intelligence. Organizations should establish proper data storage, backup, and accessibility measures to ensure data availability and security. Implementing data management systems and practices that enable easy retrieval, version control, and collaboration can streamline the data preparation process and facilitate the ongoing use of data for decision intelligence.

Conclusion:
Data acquisition and preparation form the foundation of generative AI-driven decision intelligence. By carefully defining data requirements, establishing partnerships, collecting and cleansing data, ensuring privacy and compliance, and leveraging data augmentation and enrichment techniques, organizations can build a strong data foundation. This foundation supports the training and validation of generative AI models, enabling accurate, reliable, and effective decision intelligence. With a solid data preparation strategy, organizations can unlock the full potential of generative AI-driven decision intelligence and drive data-informed, insightful decision-making.

Scalability and Computational Requirements: Empowering Generative AI for Decision Intelligence at Scale

Generative AI has emerged as a powerful technology for decision intelligence, enabling organizations to generate insights, simulate scenarios, and make data-driven decisions. However, the scalability and computational requirements associated with generative AI pose significant challenges that organizations must address to harness its full potential. In this section, we will explore the importance of scalability and computational resources in generative AI for decision intelligence and discuss strategies to overcome the associated challenges.

1. The Significance of Scalability:
Scalability is critical in generative AI-driven decision intelligence, especially when dealing with large datasets, complex models, and computationally intensive tasks. As organizations strive to analyze vast amounts of data, simulate numerous scenarios, and generate actionable insights, the ability to scale resources and processes becomes essential. Scalability enables organizations to handle increasing workloads, accommodate growing datasets, and deliver timely and responsive decision intelligence solutions.

2. Computing Power and Infrastructure:

Generative AI models often require significant computing power and infrastructure to operate efficiently. Training sophisticated models, such as deep neural networks, can be computationally demanding and may require specialized hardware, including high-performance GPUs or TPUs. Organizations must assess their computing capabilities and invest in robust infrastructure to support the computational requirements of generative AI-driven decision intelligence.

3. Cloud Computing for Scalability:

Cloud computing offers a compelling solution for addressing scalability and computational requirements. Cloud service providers offer scalable infrastructure-as-a-service (IaaS) and platform-as-a-service (PaaS) solutions, allowing organizations to dynamically scale resources based on their needs. Cloud platforms provide on-demand access to powerful computing resources, enabling organizations to handle peak workloads, reduce costs by paying only for what they use, and leverage advanced infrastructure capabilities.

4. Distributed Computing:

Distributed computing is another strategy to address scalability challenges. It involves breaking down computational tasks into smaller sub-tasks and distributing them across multiple machines or nodes in a network. By parallelizing computation, organizations can leverage the combined computing power of multiple resources to achieve faster and more scalable generative AI-driven decision intelligence. Distributed computing frameworks, such as Apache Spark or TensorFlow Distributed, facilitate the efficient distribution and coordination of computations.

5. Optimization Techniques:

Optimization techniques can help organizations optimize their generative AI models and algorithms for better scalability. Techniques such as model compression, quantization, and network pruning can reduce model size and computational requirements without significantly compromising performance. Additionally, employing efficient algorithms and data structures can improve the efficiency and speed of computations, making generative AI-driven decision intelligence more scalable.

6. AutoML and Automated Model Selection:

AutoML (Automated Machine Learning) tools and techniques can assist in addressing scalability challenges by automating the model selection and hyperparameter tuning process. These tools can explore a wide range of model architectures, configurations, and hyperparameters, searching for the best-performing models. AutoML reduces the manual effort required for model development and selection, allowing organizations to efficiently scale their generative AI-driven decision intelligence workflows.

7. Hybrid Approaches and Model Parallelism:

Hybrid approaches that combine generative AI models with traditional rule-based or expert systems can provide scalability benefits. By leveraging the strengths of both approaches, organizations can handle complex decision-making tasks efficiently. Additionally, employing model parallelism, where different parts of a generative AI model are distributed and processed across multiple devices, can enhance scalability and speed up computations.

Conclusion:

Scalability and computational requirements are vital considerations for organizations implementing generative AI for decision intelligence. By leveraging cloud computing, distributed computing, optimization techniques, AutoML, and hybrid approaches, organizations can address scalability challenges and effectively scale their generative AI-driven decision intelligence workflows. Embracing these strategies enables organizations to leverage the power of generative AI at scale, generating actionable insights, simulating scenarios, and making informed decisions that drive success in today's data-driven world.

Privacy and Security Considerations in Strategic Assessments and Scoring with Generative AI: Safeguarding Data Integrity and Confidentiality

Generative AI has brought transformative capabilities to strategic assessments and scoring, enabling organizations to gain deeper insights and make data-driven decisions. However, as organizations leverage generative AI in these processes, it is crucial to prioritize privacy and security considerations to protect the integrity and confidentiality of sensitive data. In this section, we will explore the specific privacy and security considerations that arise when using generative AI in strategic assessments and scoring and discuss strategies to ensure data protection and mitigate potential risks.

1. Privacy-Preserving Data Usage:

Generative AI models often require access to large datasets for training and generating insights. Organizations must ensure that privacy-preserving techniques are employed to protect sensitive information within the datasets. This includes implementing techniques such as differential privacy, federated learning, or secure multi-party computation, which allow organizations to analyze and generate insights from data without compromising individual privacy.

2. Secure Model Development and Deployment:

Developing and deploying generative AI models should be done securely to prevent unauthorized access or tampering. Organizations should follow secure coding practices and conduct rigorous security testing throughout the model development process. Additionally, secure deployment mechanisms, such as encryption and secure APIs, should be implemented to protect the models and the data they process from potential attacks.

3. Data Minimization and Anonymization:

To minimize privacy risks, organizations should adopt data minimization strategies by collecting and retaining only the necessary data for strategic assessments and scoring. Reducing the amount of personally identifiable information (PII) collected helps mitigate privacy concerns. Anonymization techniques, such as removing or encrypting identifiable information, can further protect individual privacy and prevent re-identification of sensitive data.

4. Consent and Transparency:

Organizations must ensure they have obtained informed consent from individuals whose data is used in generative AI-driven strategic assessments and scoring. Clear and transparent communication about the purpose of data usage, the nature of generative AI modeling, and the potential impact on privacy is essential. Organizations should provide individuals with the ability to opt out or control the usage of their data whenever feasible.

5. Data Encryption and Secure Storage:

Data encryption and secure storage practices are crucial to protect sensitive data throughout its lifecycle. Encryption should be applied both at rest and during transit to prevent unauthorized access or interception. Organizations should implement strong encryption algorithms and secure storage mechanisms, ensuring that access to the data is limited to authorized personnel only.

6. Access Controls and User Authentication:

Granular access controls and robust user authentication mechanisms should be implemented to restrict access to sensitive data and generative AI models. User roles and permissions should be defined, allowing only authorized individuals to access and modify data or interact with the generative AI system. Multi-factor authentication and strong password policies add an extra layer of security to prevent unauthorized access.

7. Regular Security Audits and Monitoring:

Regular security audits and monitoring processes are essential to identify and address any potential vulnerabilities or threats to data privacy. Organizations should conduct periodic security assessments, penetration testing, and vulnerability scans to proactively identify and mitigate security risks. Continuous monitoring of system logs and network traffic can help detect and respond to any suspicious activities in a timely manner.

8. Compliance with Regulatory Requirements:

Organizations must adhere to relevant privacy regulations, such as the General Data Protection Regulation (GDPR), the California Consumer Privacy Act (CCPA), or industry-specific regulations. Compliance with these regulations ensures that privacy rights are protected and individuals' data is handled in a lawful and responsible manner. Organizations should stay updated with evolving privacy regulations and adjust their practices accordingly.

Conclusion:

Privacy and security considerations are of utmost importance when leveraging generative AI in strategic assessments and scoring. By implementing privacy-preserving techniques, securing model development and deployment, practicing data minimization and anonymization, obtaining informed consent, and enforcing stringent access controls, organizations can ensure the confidentiality, integrity, and privacy of sensitive data. By prioritizing privacy and security, organizations can confidently leverage generative AI to drive strategic decision-making while maintaining the trust and confidence of stakeholders.

Organizational Readiness and Change Management for Generative AI Adoption: Navigating the Path to Success

Generative AI has the potential to revolutionize industries, drive innovation, and unlock new possibilities for organizations. However, successfully adopting generative AI requires more than just implementing the technology—it necessitates careful consideration of organizational readiness and effective change management strategies. In this section, we will explore the importance of organizational readiness and change management in the adoption of generative AI and discuss strategies to navigate the path to success.

1. Assessing Organizational Readiness:
Before embarking on the adoption of generative AI, organizations must assess their readiness for change. This involves evaluating factors such as the organization's culture, leadership support, technical capabilities, and data infrastructure. Assessing readiness helps identify potential barriers and gaps that need to be addressed to ensure a smooth transition.

2. Establishing a Clear Vision and Objectives:
Organizations should define a clear vision and objectives for generative AI adoption. This includes understanding the specific goals, benefits, and value proposition of using generative AI. The vision should be communicated to all stakeholders, creating a shared understanding of how generative AI aligns with the organization's strategic objectives and contributes to its success.

3. Building Cross-Functional Teams:
Generative AI adoption requires collaboration and coordination across different functions and teams within the organization. Building cross-functional teams helps ensure diverse perspectives, expertise, and knowledge are integrated into the adoption process. These teams can include representatives from IT, data science, business units, legal, and other relevant departments.

4. Training and Upskilling:
Generative AI adoption often requires new skills and capabilities within the organization. Providing training and upskilling opportunities for employees is crucial to ensure they have the necessary knowledge and expertise to effectively work with generative AI. This can involve technical training in AI technologies, data analytics, and machine learning, as well as developing a broader

understanding of the implications and potential applications of generative AI.

5. Change Management and Communication:
Effective change management is essential for successful generative AI adoption. Organizations should develop a change management plan that includes clear communication strategies, stakeholder engagement, and addressing potential resistance to change. Transparent and frequent communication helps create awareness, gain buy-in, and alleviate concerns about the impact of generative AI on job roles and responsibilities.

6. Piloting and Iterative Approach:
Piloting generative AI initiatives allows organizations to test the technology in a controlled environment before scaling up. Pilots provide valuable insights, identify challenges, and enable organizations to iterate and refine their approach. Taking an iterative approach allows for continuous learning, adaptation, and improvement, ensuring that generative AI adoption aligns with evolving organizational needs.

7. Data Governance and Ethical Considerations:
Generative AI relies heavily on data, making data governance and ethical considerations crucial. Organizations must establish robust data governance practices, including data quality control, privacy protection, and compliance with regulatory requirements. Ethical considerations, such as bias mitigation, transparency, and accountability, should also be addressed to ensure responsible and ethical use of generative AI.

8. Collaboration with External Partners:
Collaborating with external partners, such as AI experts, consultants, or technology vendors, can provide valuable support and expertise during generative AI adoption. External partners can offer guidance on technical implementation, best practices, and industry insights, helping organizations navigate the complexities of generative AI adoption more effectively.

9. Continuous Monitoring and Evaluation:
Once generative AI is adopted, continuous monitoring and evaluation are crucial to assess its impact, identify areas for improvement, and ensure the technology delivers the expected outcomes. Regular monitoring allows organizations to track performance, address emerging challenges, and make informed decisions regarding scaling, refinement, or potential course corrections.

Conclusion:
Generative AI adoption requires careful attention to organizational readiness and effective change management strategies. By assessing readiness, establishing a clear vision, building cross-functional teams, providing training and upskilling, implementing effective change management processes, piloting, addressing data governance and ethical considerations, collaborating with external partners, and continuously monitoring progress, organizations can navigate the path to

successful generative AI adoption. With the right organizational readiness and change management approach, organizations can harness the full potential of generative AI to drive innovation and achieve transformative outcomes.

Managing a major project in these challenging times requires IMAGINATION, the FIRST thing I'd like you to imagine is that you have a team to help you.

Figure 9

Chapter 8: Imagining the Future: Decision Intelligence and Generative AI

The future of decision-making is being reshaped by the convergence of decision intelligence and generative AI. As organizations strive to navigate complex environments, make data-driven choices, and unlock new insights, the combination of these two powerful technologies holds tremendous promise. In this section, we will explore the potential of decision intelligence and generative AI, envisioning the transformative possibilities they offer for the future of decision-making.

1. Decision Intelligence: Enhancing Decision-Making Capabilities

Decision intelligence is the field that combines human decision-making with advanced analytics and AI techniques to improve the quality of decisions. By leveraging data, algorithms, and human expertise, decision intelligence enhances decision-making capabilities across various domains. It integrates data-driven insights, predictive models, optimization algorithms, and visualization techniques to enable organizations to make more informed, evidence-based decisions.

2. Generative AI: Creating and Innovating

Generative AI, on the other hand, focuses on creating new content, models, or solutions through machine learning techniques. It goes beyond traditional AI approaches that rely on pre-existing data and rules. Generative AI models can generate novel outputs, such as images, texts, or even entire scenarios, based on the patterns and information learned during training. This technology has shown immense potential in various creative fields, including art, design, and content creation.

3. Convergence of Decision Intelligence and Generative AI

The convergence of decision intelligence and generative AI opens up exciting possibilities for the future of decision-making. By leveraging generative AI within decision intelligence frameworks, organizations can not only gain deeper insights from existing data but also generate new data, simulate scenarios, and explore innovative solutions that were previously unimagined. Generative AI complements decision intelligence by expanding the range of possibilities and empowering organizations to think beyond traditional boundaries.

4. Enhanced Scenario Analysis and Planning

Generative AI enables organizations to create diverse scenarios and simulate potential outcomes, contributing to more robust scenario analysis and planning. Decision intelligence frameworks can leverage generative AI to generate alternative scenarios, test the impact of different variables, and explore the consequences of various decisions. This provides organizations with a broader perspective and a more comprehensive understanding of the potential risks and opportunities associated with different strategies.

5. Data Synthesis and Augmentation

Generative AI can assist in data synthesis and augmentation, overcoming limitations caused by data scarcity or gaps. Decision intelligence relies heavily on data, and generative AI can generate synthetic data to supplement existing datasets, enabling more accurate and comprehensive analyses. By synthesizing data, generative AI can help organizations overcome data limitations, resulting in more reliable and robust decision intelligence models.

6. Unleashing Creativity and Innovation

The integration of generative AI into decision intelligence frameworks unlocks new possibilities for creativity and innovation in decision-making. Generative AI models can generate novel ideas, designs, or solutions that challenge conventional thinking and inspire organizations to explore new avenues. By augmenting human creativity with generative AI, decision intelligence becomes a catalyst for innovation, fostering a culture of continuous improvement and exploration.

7. Ethical and Responsible Application

As decision intelligence and generative AI evolve, it is essential to prioritize ethical and responsible application. Organizations must consider the potential biases, fairness, and transparency of the generative AI models used within decision intelligence frameworks. Ethical guidelines, accountability, and continuous monitoring are crucial to ensure that generative AI-driven decision-making aligns with societal values and respects individual rights.

Conclusion: A New Frontier in Decision-Making

The convergence of decision intelligence and generative AI represents a new frontier in decision-making, one that holds immense potential for organizations across industries. By combining the analytical power of decision intelligence with the creative capabilities of generative AI, organizations can gain deeper insights, explore innovative solutions, and make informed decisions that drive success. Embracing this future requires organizations to foster a culture of collaboration, invest in advanced technologies, and prioritize ethical considerations. As decision intelligence and generative AI continue to evolve, the future of decision-making will undoubtedly be shaped by their synergistic capabilities, unlocking new horizons of possibilities.

Exploring Emerging Trends and Advancements in Generative AI for Strategy Assessments

Generative AI, a branch of artificial intelligence that focuses on creating new content, models, or solutions, is rapidly advancing and transforming various industries. In the realm of strategy assessments, generative AI offers innovative approaches to analyze data, simulate scenarios, and generate insights that can inform and enhance strategic decision-making. In this section, we will explore some of the emerging trends and advancements in generative AI that are shaping strategy assessments.

1. Deep Generative Models:

Deep generative models, such as generative adversarial networks (GANs) and variational autoencoders (VAEs), are at the forefront of generative AI advancements. GANs can generate highly realistic synthetic data by training a generator network to generate samples that resemble real data, while a discriminator network learns to distinguish between real and synthetic data. VAEs, on the other hand, learn the underlying distribution of data and can generate new samples based on that distribution. These models enable organizations to explore a wide range of possibilities and generate synthetic data that can augment strategy assessments.

2. Text and Language Generation:

Advancements in generative AI have led to significant progress in text and language generation. Natural Language Processing (NLP) models, such as OpenAI's GPT-3, have demonstrated remarkable capabilities in generating coherent and contextually relevant text. In strategy assessments, these models can generate written reports, summaries, or insights based on the analysis of complex datasets. They can also aid in generating natural language responses for chatbots or virtual assistants, facilitating interactive and dynamic strategy assessments.

3. Image and Visual Content Generation:

Generative AI techniques have also made significant strides in generating visual content, including images, designs, and artwork. Style transfer algorithms can transform images to resemble famous art styles or create entirely new and visually appealing designs. These capabilities have implications for strategy assessments, as organizations can leverage generative AI to generate visual representations of data, visualize scenarios, or create engaging visual content to communicate insights effectively.

4. Reinforcement Learning for Strategy Optimization:

Reinforcement learning, a subset of machine learning, focuses on training agents to learn optimal decision-making strategies through trial and error. In strategy assessments, reinforcement learning techniques can be applied to optimize decision-making processes by identifying the most effective strategies or policies. Generative AI models trained through reinforcement learning can generate recommendations or simulations to guide organizations in selecting the best strategic initiatives.

5. Hybrid Approaches and Cross-Domain Applications:

Hybrid approaches that combine generative AI with other AI techniques, such as reinforcement learning or rule-based systems, are gaining traction. These approaches leverage the strengths of different AI methods to tackle complex strategy assessment challenges. Additionally, generative AI is finding applications in various domains, including finance, healthcare, marketing, and supply chain management. Its ability to generate synthetic data, simulate scenarios, and generate insights makes it a versatile tool for strategy assessments across industries.

6. Explainability and Interpretability:

As generative AI models become more sophisticated, there is an increasing focus on explainability and interpretability. Understanding how and why a generative AI model generates specific outputs is crucial for building trust and ensuring transparency in strategy assessments. Advancements in explainable AI techniques, such as attention mechanisms or feature attribution methods, are enabling organizations to gain insights into the decision-making process of generative AI models.

7. Ethical Considerations and Bias Mitigation:

Ethical considerations and bias mitigation are integral to responsible and fair strategy assessments using generative AI. Organizations must be vigilant in addressing potential biases in the training data or algorithms used in generative AI models. This includes careful selection of training data, continuous monitoring for biases, and implementing mitigation strategies to ensure fair and unbiased assessments.

Conclusion:

Generative AI is poised to revolutionize strategy assessments by offering novel ways to analyze data, simulate scenarios, and generate insights. Emerging trends and advancements in generative AI, such as deep generative models, text and language generation, image and visual content generation, reinforcement learning, hybrid approaches, explainability, and ethical considerations, are shaping the future of strategy assessments. By embracing these advancements and considering their implications, organizations can harness the transformative power of generative AI and drive strategic decision-making to new heights of effectiveness and innovation.

Envisioning the Future of Decision Intelligence with Generative AI

The future of decision intelligence is rapidly evolving, driven by the integration of generative AI. Generative AI, with its ability to create new content, models, or solutions, holds immense potential to transform decision-making processes and revolutionize the field of decision intelligence. In this section, we will explore the exciting possibilities and envision the future of decision intelligence as it converges with generative AI.

1. Enhanced Data Analysis and Insight Generation:
Generative AI can unlock deeper insights from complex datasets by generating synthetic data, augmenting existing data, or simulating scenarios. In the future, decision intelligence powered by generative AI can revolutionize data analysis by providing novel perspectives and generating valuable insights that may have been previously inaccessible. This enables organizations to make more informed and data-driven decisions.

2. Scenario Generation and Optimization:
Generative AI has the potential to transform scenario generation and optimization within decision intelligence. It can generate a vast array of scenarios based on different variables, helping decision-makers explore a wide range of possibilities. By leveraging generative AI, decision intelligence can identify optimal strategies, optimize resource allocation, and guide organizations in making informed choices in complex and uncertain environments.

3. Creative Solution Generation:
Generative AI excels in creative content generation, such as images, designs, or text. In decision intelligence, this creative potential can be harnessed to generate innovative solutions and alternatives. By combining human expertise with generative AI capabilities, decision intelligence can push the boundaries of creativity, enabling organizations to explore unconventional strategies and uncover novel approaches to problem-solving.

4. Real-time Decision Support:
The integration of generative AI with decision intelligence paves the way for real-time decision support. As generative AI models learn and adapt from new data, decision intelligence systems can provide up-to-date insights and recommendations. Real-time decision support empowers organizations to respond swiftly to dynamic market conditions, changing customer preferences, or emerging trends, enabling agile and proactive decision-making.

5. Personalized Decision Intelligence:
Generative AI can be leveraged to create personalized decision intelligence experiences. By learning from individual preferences, behavior, and historical data, generative AI models can provide tailored recommendations and insights to decision-makers. This personalized approach

enhances the relevance and effectiveness of decision intelligence, catering to the specific needs and objectives of each stakeholder.

6. Augmented Collaboration:
Generative AI has the potential to facilitate augmented collaboration within decision intelligence processes. By generating synthetic data or scenarios, generative AI models can serve as a catalyst for collaboration, stimulating discussions, and fostering diverse perspectives. Decision-makers can interact with generative AI models to explore different scenarios, test assumptions, and collectively make informed decisions.

7. Ethical and Responsible Decision Intelligence:
As generative AI becomes more integrated into decision intelligence, ethical and responsible decision-making becomes paramount. Organizations must address potential biases, interpretability, and transparency issues associated with generative AI models. The future of decision intelligence with generative AI will involve developing robust ethical frameworks, adhering to regulatory standards, and ensuring that decision-making processes are fair, accountable, and aligned with societal values.

Conclusion:
The future of decision intelligence is bright and promising with the integration of generative AI. By leveraging generative AI capabilities, decision intelligence can unlock deeper insights, enable creative solution generation, provide real-time decision support, foster collaboration, and ensure ethical and responsible decision-making. As organizations embrace the potential of generative AI within decision intelligence, they can navigate complex challenges, seize new opportunities, and make more informed, innovative decisions that drive success in the ever-evolving business landscape.

Collaborative Decision-Making between Human Experts and Generative AI Algorithms: Unleashing the Power of Synergy

In today's rapidly evolving world, the collaboration between human experts and artificial intelligence (AI) algorithms has the potential to revolutionize decision-making processes. Specifically, the synergy between human expertise and generative AI algorithms has emerged as a powerful approach to tackle complex challenges, generate innovative solutions, and make more informed decisions. In this section, we will explore the concept of collaborative decision-making between human experts and generative AI algorithms, highlighting the benefits, challenges, and key considerations for successful implementation.

1. Leveraging Human Expertise:
Human expertise brings unique insights, intuition, and contextual understanding to the decision-making process. Experts possess domain knowledge, experience, and the ability to consider various factors beyond data inputs. By involving human experts, collaborative decision-making combines

their deep understanding of complex problems with the analytical capabilities of generative AI algorithms, resulting in a holistic and well-informed decision-making process.

2. Harnessing Generative AI Algorithms:

Generative AI algorithms, such as generative adversarial networks (GANs) or variational autoencoders (VAEs), excel at generating new content, patterns, or scenarios based on learned patterns from large datasets. These algorithms can augment human decision-making by simulating various scenarios, generating alternative solutions, and identifying patterns that might not be immediately apparent to human experts. Generative AI algorithms complement human expertise by providing novel perspectives and expanding the solution space.

3. Uncovering Hidden Insights:

The collaboration between human experts and generative AI algorithms can uncover hidden insights and identify novel patterns that may not be apparent through traditional analysis alone. Generative AI algorithms have the ability to process large volumes of data, identify complex correlations, and generate synthetic data for further analysis. By combining these capabilities with human expertise, decision-makers can gain deeper insights, discover non-obvious connections, and make more informed decisions based on a comprehensive understanding of the problem space.

4. Augmenting Creativity and Innovation:

Collaborative decision-making with generative AI algorithms can foster creativity and innovation. The algorithms can generate a wide range of ideas, designs, or solutions that challenge conventional thinking and provide fresh perspectives. Human experts can then evaluate, refine, and further develop these generated ideas, leveraging their domain knowledge and expertise. The synergy between human creativity and generative AI algorithms' capacity for exploration can lead to breakthrough solutions and drive innovation.

5. Addressing Limitations and Ethical Considerations:

While the collaboration between human experts and generative AI algorithms offers numerous benefits, it is essential to acknowledge and address potential limitations and ethical considerations. Generative AI algorithms may be prone to biases present in training data, requiring careful monitoring and mitigation strategies. Transparency, interpretability, and accountability should also be ensured to understand the decision-making process and mitigate potential risks.

6. Building Trust and Effective Communication:

Successful collaborative decision-making relies on building trust and effective communication between human experts and generative AI algorithms. Human experts need to understand the capabilities and limitations of the generative AI algorithms, while the algorithms should be transparently presented to ensure human experts can provide critical feedback and validation. Continuous communication, feedback loops, and shared decision-making frameworks are essential for fostering a productive and mutually beneficial collaboration.

7. Continuous Learning and Adaptation:

Collaborative decision-making between human experts and generative AI algorithms should be seen as an iterative and evolving process. Learning from both successes and failures, organizations can refine their approach, update models, and adapt decision-making strategies. This continuous learning cycle enhances the effectiveness of the collaboration and ensures the evolution of decision-making capabilities over time.

Conclusion:

Collaborative decision-making between human experts and generative AI algorithms represents a powerful paradigm for addressing complex challenges and driving innovation. By leveraging the unique strengths of both human expertise and generative AI algorithms, organizations can unlock new insights, generate creative solutions, and make informed decisions that harness the full potential of AI while benefiting from human judgment and domain knowledge. Embracing the synergy between human and artificial intelligence is key to shaping the future of decision-making in a rapidly advancing world.

Integrating Generative AI with Other Transformative Technologies for Enhanced Decision Intelligence: Unlocking New Possibilities

In today's era of digital transformation, organizations are continuously seeking ways to leverage cutting-edge technologies to enhance their decision intelligence capabilities. One particularly powerful approach is integrating generative AI with other transformative technologies. By combining generative AI with emerging technologies like machine learning, big data analytics, Internet of Things (IoT), and natural language processing (NLP), organizations can unlock new possibilities for decision intelligence. In this section, we will explore the benefits, applications, and key considerations of integrating generative AI with other transformative technologies to achieve enhanced decision intelligence.

1. Machine Learning:

Machine learning techniques, including deep learning, can be integrated with generative AI to improve decision intelligence. By combining the power of generative AI in generating new content or scenarios with the analytical capabilities of machine learning, organizations can gain deeper insights, predict outcomes, and make data-driven decisions. Machine learning models can be trained on generative AI-generated data or combined with generative models to improve prediction accuracy and optimize decision-making processes.

2. Big Data Analytics:

Integrating generative AI with big data analytics allows organizations to process and analyze large volumes of data to uncover hidden patterns, trends, and correlations. Generative AI models can generate synthetic data that augments existing datasets, allowing for more comprehensive analysis and better decision-making. By combining generative AI with big data analytics, organizations can gain a holistic view of their data landscape and derive meaningful insights to inform strategic decisions.

3. Internet of Things (IoT):

The integration of generative AI with IoT opens up new opportunities for decision intelligence. IoT devices generate vast amounts of data, which can be leveraged by generative AI algorithms to simulate scenarios, predict outcomes, or generate insights. For example, generative AI models can analyze IoT data to identify patterns and anomalies, optimize resource allocation, or predict equipment failures, enabling proactive decision-making and improving operational efficiency.

4. Natural Language Processing (NLP):

Generative AI can be integrated with NLP techniques to enhance decision intelligence in textual data analysis and understanding. NLP algorithms enable the extraction of insights from unstructured text, such as customer reviews, social media posts, or reports. By combining generative AI with NLP, organizations can generate summaries, sentiment analyses, or context-aware recommendations, enabling decision-makers to gain deeper understanding and make informed decisions based on textual data.

5. Robotics and Automation:

Integrating generative AI with robotics and automation technologies can enhance decision intelligence in areas such as process optimization and autonomous systems. Generative AI algorithms can generate simulated scenarios or generate robot behavior, enabling organizations to evaluate different strategies or automate decision-making processes. This integration can lead to increased efficiency, reduced errors, and improved decision-making in various industries.

6. Virtual and Augmented Reality (VR/AR):

Virtual and augmented reality technologies can be combined with generative AI to create immersive decision intelligence experiences. By visualizing generative AI-generated data or scenarios through VR/AR, decision-makers can explore, interact, and gain deeper insights into complex problems. This integration facilitates a more intuitive and immersive decision-making process, empowering stakeholders to make more informed and effective decisions.

Considerations for Successful Integration

While integrating generative AI with other transformative technologies offers tremendous potential for enhanced decision intelligence, there are key considerations to ensure success:

1. Data Quality and Ethics:

High-quality data is essential for effective integration. Organizations must ensure data integrity, address biases, and comply with ethical guidelines to prevent unintended consequences or unfair decision-making outcomes.

2. Interdisciplinary Collaboration:

Successful integration requires collaboration among experts from various domains, including AI, data science, domain specialists, and business stakeholders. Interdisciplinary collaboration fosters the effective integration of generative AI with other technologies, ensuring a holistic and well-informed decision intelligence approach.

3. Scalability and Infrastructure:

Integrating generative AI with other technologies may require robust infrastructure and scalable systems to handle the computational requirements and data processing capabilities. Organizations must assess and invest in the necessary infrastructure to support seamless integration.

4. Change Management and Training:

Integrating transformative technologies requires change management efforts, including training employees on new tools, processes, and methodologies. Building a culture that embraces technology integration and provides continuous learning opportunities is crucial for successful implementation.

Conclusion:

Integrating generative AI with other transformative technologies has the potential to revolutionize decision intelligence by providing deeper insights, predictive capabilities, and enhanced data analysis. By harnessing the power of generative AI alongside technologies like machine learning, big data analytics, IoT, NLP, robotics, and VR/AR, organizations can make more informed, data-driven decisions, optimize processes, and unlock new opportunities. The successful integration of these transformative technologies requires a strategic approach, interdisciplinary collaboration, and a commitment to ethical considerations, ultimately leading to enhanced decision intelligence and organizational success in today's dynamic business landscape.

Conclusion of Section 1 – The Why:

As we conclude our exploration of decision intelligence with generative AI for assessing, scoring, and prioritizing enterprise strategy initiatives, we recognize the transformative potential of this fusion for organizations seeking to navigate complexity and make strategic choices. By leveraging generative AI to illuminate pathways, enterprises can make more informed and impactful decisions, ultimately driving success and competitive advantage. It is crucial to navigate ethical considerations and implementation.

Illuminating Pathways

Section 2 – The How:
Generative AI for Enterprise Decision
Intelligence

Introduction

Harnessing Innovation and Insight for Optimal Business Choices

Introduction

In an increasingly competitive and data-rich business landscape, making informed decisions is crucial for enterprise success. The emergence of Generative AI has paved the way for a new era of Decision Intelligence, empowering organizations to leverage the power of artificial intelligence to drive innovation, optimize operations, and navigate complex challenges. This book explores the intersection of Generative AI and Enterprise Decision Intelligence, offering insights, strategies, and practical applications for harnessing the full potential of this transformative technology.

"I don't understand the technology either, but I
feel alright as long as there's still a plug to pull."

Figure 10

Chapter 9: The Foundations of Enterprise Decision Intelligence

Understanding Enterprise Decision Intelligence Concepts, Principles, and Benefits.

In today's complex and rapidly evolving business landscape, making informed decisions is crucial for the success and growth of enterprises. Enterprises must navigate a multitude of factors, such as market conditions, customer preferences, operational constraints, and competitive dynamics, to make the right choices. To support this process, a relatively new field of study has emerged called Enterprise Decision Intelligence (EDI). EDI combines various disciplines, including artificial intelligence, data analytics, and decision science, to provide a comprehensive framework for understanding and improving decision-making within organizations. In this section, we will explore the concepts, principles, and benefits of Enterprise Decision Intelligence.

Concepts of Enterprise Decision Intelligence

1. Data-Driven Decision-Making: EDI emphasizes the importance of using data to drive decision-making processes. By leveraging large volumes of structured and unstructured data, organizations can gain valuable insights into their operations, customers, and markets. Data analytics techniques, such as statistical analysis, machine learnins from the data. These insights serve as the foundation for informed decision-making.

2. Decision Support Systems: EDI relies on decision support systems (DSS) to facilitate and enhance decision-making processes. DSS are computer-based tools that provide interactive and analytical capabilities to aid decision-makers. They integrate data from various sources, perform sophisticated analyses, and present the information in a user-friendly manner. DSS enable decision-makers to evaluate different scenarios, assess risks, and explore alternative options, leading to more effective and efficient decisions.

3. Cognitive Computing: Another key concept in EDI is cognitive computing, which involves the use of advanced technologies, such as natural language processing, machine learning, and knowledge representation, to mimic human thought processes. Cognitive computing systems can understand and interpret unstructured data, learn from past decisions, and provide intelligent recommendations to decision-makers. These systems enhance human cognition, enabling organizations to make better decisions in complex and uncertain environments.

Principles of Enterprise Decision Intelligence

1. Holistic Decision-Making: EDI emphasizes the importance of taking a holistic approach to decision-making. Instead of making isolated choices, EDI encourages decision-makers to consider the interdependencies and trade-offs across different aspects of the enterprise. This includes understanding the impact of decisions on financial performance, operational efficiency, customer satisfaction, and long-term sustainability. By considering the broader implications, organizations can make decisions that align with their strategic objectives and optimize overall performance.

2. Iterative and Adaptive Decision-Making: EDI recognizes that decision-making is an iterative and adaptive process. As new data becomes available or circumstances change, decisions may need to be revisited and adjusted. EDI promotes continuous learning and improvement through feedback loops and real-time monitoring of decision outcomes. This allows organizations to make timely adjustments and course corrections, increasing agility and responsiveness in a rapidly evolving business environment.

3. Human-Centric Design: While EDI leverages advanced technologies and analytics, it emphasizes the importance of human-centric design. EDI systems are designed to support and augment human decision-making rather than replacing it. The goal is to empower decision-makers with timely and relevant information, insights, and recommendations, enabling them to make well-informed choices based on their expertise and judgment. Human-machine collaboration is at the core of EDI, combining the strengths of both to achieve superior decision outcomes.

Benefits of Enterprise Decision Intelligence

1. Improved Decision Quality: By leveraging data and advanced analytics, EDI enables organizations to make more informed decisions with higher accuracy and reliability. The insights generated by EDI systems help decision-makers understand the impact of their choices and evaluate various scenarios before making a final decision. This leads to improved decision quality and reduces the likelihood of making costly mistakes.

2. Enhanced Efficiency and Productivity: EDI streamlines decision-making processes by automating routine tasks, reducing manual effort, and providing decision-makers with relevant information at their fingertips. By eliminating information silos and facilitating collaboration, EDI improves the efficiency and productivity of decision-making across the enterprise. Decision-makers can focus on strategic and value-added activities, leading to faster and more effective outcomes.

3. Competitive Advantage: In today's data-driven economy, organizations that effectively leverage data and analytics have a competitive edge. EDI enables enterprises to unlock the value of their data assets and gain deeper insights into their business operations, customers, and markets. This helps identify new opportunities, anticipate market trends, and make proactive decisions, giving organizations a competitive advantage in a rapidly changing business landscape.

4. Risk Mitigation: Effective decision-making requires understanding and managing risks. EDI provides organizations with the ability to assess and mitigate risks by incorporating risk models, scenario analysis, and predictive analytics into decision-making processes. By considering potential risks and uncertainties, organizations can make decisions that are more robust and resilient, reducing the likelihood of negative outcomes and enhancing long-term sustainability.

In conclusion, Enterprise Decision Intelligence (EDI) is a multidisciplinary field that combines data analytics, decision science, and advanced technologies to support informed decision-making within organizations. By leveraging data, cognitive computing, and decision support systems, EDI enables organizations to make better decisions that align with their strategic objectives and optimize performance. The concepts, principles, and benefits of EDI emphasize the importance of holistic decision-making, iterative learning, and human-machine collaboration. With EDI, organizations can improve decision quality, enhance efficiency and productivity, gain a competitive advantage, and mitigate risks, ultimately leading to better business outcomes in an increasingly complex and dynamic business environment.

The Role Of Generative AI In Decision-Making: From Data Analytics To Synthetic Data Generation

In recent years, the field of artificial intelligence (AI) has witnessed significant advancements, particularly in the area of generative AI. Generative AI refers to a class of AI techniques that involve the creation and generation of new content, such as images, text, and even data. This technology has been revolutionizing decision-making processes across various industries, enabling organizations to leverage data analytics and synthetic data generation to make more informed and effective decisions. In this section, we will explore the role of generative AI in decision-making, specifically focusing on its applications in data analytics and synthetic data generation.

Data Analytics

Data analytics is a fundamental component of decision-making in today's data-driven world. It involves extracting meaningful insights and patterns from vast amounts of data to support informed decision-making. Generative AI techniques, such as generative adversarial networks (GANs) and variational autoencoders (VAEs), have emerged as powerful tools for data analytics.

1. Data Augmentation: Generative AI can be used to augment existing datasets by generating synthetic data that closely resembles the original data. This process helps to overcome limitations such as data scarcity and imbalance, thereby enhancing the quality and diversity of the training data. By generating additional data points, organizations can improve the accuracy and robustness of their predictive models and analytical algorithms, leading to more reliable decision outcomes.

2. Anomaly Detection: Generative AI models can be trained on normal data patterns and used to detect anomalies or outliers in new data. By comparing the generated data with the real data, organizations can identify patterns that deviate from the norm and potentially indicate fraudulent activities, system failures, or other abnormal events. This helps organizations proactively address issues and make timely decisions to mitigate risks.

3. Exploratory Data Analysis: Generative AI techniques enable organizations to explore and visualize data in novel ways. By generating new samples based on the existing data distribution, decision-makers can gain a deeper understanding of the underlying data patterns and relationships. This allows for more comprehensive exploratory data analysis, uncovering insights that may have been overlooked with traditional methods.

Synthetic Data Generation

In addition to data analytics, generative AI plays a crucial role in synthetic data generation. Synthetic data refers to artificially generated data that resembles real-world data but does not contain any personally identifiable information or sensitive content. Synthetic data generation offers several benefits for decision-making processes:

1. Privacy Protection: In many industries, privacy regulations and data protection concerns restrict the usage and sharing of sensitive data. Generative AI can generate synthetic data that retains the statistical properties and patterns of the original data while ensuring privacy and anonymity. Decision-makers can use this synthetic data for analysis and modeling without compromising individual privacy, facilitating compliance with privacy regulations.

2. Simulation and Scenario Analysis: Generative AI enables the creation of synthetic datasets that mimic different scenarios or hypothetical situations. Decision-makers can use these datasets to simulate various scenarios and assess the potential outcomes and impacts of their decisions. This helps organizations evaluate different strategies, anticipate risks, and make more informed decisions, especially in complex and uncertain environments.

3. Data Diversity and Generalization: Real-world datasets may be limited in terms of diversity, making it challenging to build models that generalize well. Generative AI can be used to generate synthetic data that spans a broader range of variations and scenarios. This allows decision-makers to train models on more comprehensive datasets, leading to improved generalization and better decision-making outcomes.

However, it is important to note that while generative AI offers significant opportunities, it also raises ethical considerations. Care must be taken to ensure the responsible and ethical use of generative AI technologies to prevent any unintended biases or unfairness in decision-making processes.

In conclusion, generative AI plays a vital role in decision-making processes by leveraging data analytics and synthetic data generation. By augmenting datasets, detecting anomalies, and enabling exploratory data analysis, generative AI enhances the quality and reliability of decision outcomes. Additionally, synthetic data generation addresses privacy concerns, enables simulation and scenario analysis, and improves data diversity and generalization. As organizations continue to embrace AI technologies, the role of generative AI in decision-making will undoubtedly expand, empowering decision-makers with powerful tools to navigate the complexities of the modern business landscape.

Human-AI collaboration: Augmenting Human Expertise with Generative Models

As artificial intelligence (AI) technologies continue to advance, there is an increasing recognition of the potential for collaboration between humans and AI systems. Rather than viewing AI as a replacement for human expertise, organizations are embracing the concept of Human-AI collaboration, where AI is used to augment and enhance human capabilities. One powerful application of this collaboration is the use of generative models, which can generate new content based on existing data. In this section, we will explore the concept of Human-AI collaboration and how generative models can augment human expertise.

Understanding Human-AI Collaboration:
Human-AI collaboration involves harnessing the strengths of both humans and AI systems to

achieve superior outcomes. While AI systems excel at processing vast amounts of data, recognizing patterns, and making predictions, they lack the intuition, creativity, and contextual understanding that humans possess. By combining the analytical power of AI with human expertise, organizations can leverage the complementary strengths of both to solve complex problems and make informed decisions.

Augmenting Human Expertise with Generative Models:

Generative models, such as generative adversarial networks (GANs) and variational autoencoders (VAEs), are AI techniques that can generate new content based on patterns and structures learned from existing data. These models have shown remarkable capabilities in various domains, including image synthesis, text generation, and music composition. When used in collaboration with human experts, generative models can provide several benefits:

1. Idea Generation and Creativity: Human experts often rely on their domain knowledge and experience to generate ideas and solutions. Generative models can assist in the creative process by generating new possibilities based on existing patterns and examples. By presenting alternative options and novel perspectives, generative models can inspire human experts and expand the range of potential solutions.

2. Design and Optimization: In fields such as product design, architecture, or manufacturing, generative models can assist human experts in the design and optimization process. By exploring different design variations and automatically generating alternatives, these models can help identify optimal solutions based on specific constraints and objectives. Human experts can then refine and further improve upon the generated designs, leading to more efficient and innovative outcomes.

3. Decision Support and Scenario Analysis: When faced with complex decision-making scenarios, generative models can assist human experts by generating synthetic data or simulating various scenarios. By exploring different possibilities and assessing potential outcomes, human experts can make more informed decisions based on a broader range of information. Generative models facilitate scenario analysis, risk assessment, and sensitivity analysis, empowering human decision-makers with valuable insights.

4. Expertise Transfer and Training: Generative models can also be used to transfer and preserve human expertise. By learning from vast amounts of data and examples provided by human experts, generative models can capture and encode their knowledge, style, and preferences. This can be particularly useful in fields where expert knowledge is scarce or where human experts are retiring or transitioning. The generative models can then be used to assist and train new professionals, ensuring the preservation and continuity of valuable expertise.

Ethical Considerations and Challenges:

While the collaboration between humans and generative models offers significant opportunities, it is essential to address ethical considerations and challenges. Some key considerations include:

1. Bias and Fairness: Generative models learn from the data they are trained on, which can introduce biases present in the training data. It is crucial to ensure that the training data is diverse, representative, and free from discriminatory biases to avoid perpetuating or amplifying unfairness.

2. Accountability and Responsibility: In collaborative decision-making processes, it is important to define clear roles and responsibilities for both humans and generative models. Humans should maintain ultimate accountability for decisions and actively oversee the outputs generated by AI systems to avoid blindly following the suggestions of the models.

3. Transparency and Explainability: Generative models are often complex, black-box systems, making it challenging to understand how they generate specific outputs. Ensuring transparency and explainability is crucial, particularly in high-stakes domains where decision-making must be justified and understood.

Conclusion:

Human-AI collaboration, particularly through the use of generative models, offers tremendous potential for augmenting human expertise and enhancing decision-making processes. By leveraging the strengths of both humans and AI systems, organizations can benefit from increased creativity, improved design and optimization, enhanced decision support, and efficient expertise transfer. However, careful consideration of ethical implications, transparency, and accountability is necessary to ensure the responsible and effective use of generative models in collaborative settings. Ultimately, the synergy between human expertise and generative models paves the way for innovative and impactful outcomes across various domains.

The impact of Enterprise Decision Intelligence on organizational performance and agility

In today's fast-paced and competitive business landscape, organizations are constantly seeking ways to improve their performance and increase their agility. One powerful approach that has emerged in recent years is Enterprise Decision Intelligence (EDI). EDI combines advanced technologies, data analytics, and decision science to enhance decision-making processes within organizations. In this section, we will explore the impact of Enterprise Decision Intelligence on organizational performance and agility.

1. Improved Decision Quality: One of the primary impacts of EDI is the improvement in decision quality. By leveraging data analytics, AI algorithms, and decision support systems, EDI enables organizations to make more informed and data-driven decisions. It helps decision-makers understand complex relationships, identify patterns, and evaluate various scenarios before making a choice. The result is higher-quality decisions that align with organizational goals, leading to

improved performance
ı

2. Enhanced Opertional Efficiency: EDI optimizes operational efficiency by streamlining decision-making pro[esses. By automating routine tasks, providing real-time data insights, and facilitating collaboration, EDI eliminates inefficiencies and reduces manual effort. Decision-makers can access relevant information and analysis quickly, enabling faster and more effective decision-making. This improved efficiency translates into cost savings, resource optimization, and streamlined operations.

3. Proactive Risk Management: EDI enables organizations to proactively manage risks and uncertainties. By integrating risk models, predictive analytics, and scenario analysis into decision-making processes, EDI helps identify potential risks and assess their impact on the organization. Decision-makers can evaluate risk-reward trade-offs, implement mitigation strategies, and make informed decisions that minimize exposure to risks. This proactive approach to risk management enhances organizational resilience and agility.

4. Increased Agility and Adaptability: Agility is crucial for organizations to respond and adapt to changing market conditions and customer demands. EDI supports agility by providing real-time insights, enabling rapid decision-making, and facilitating course corrections. By continuously monitoring data, market trends, and performance metrics, organizations can quickly identify opportunities and threats, make timely adjustments, and pivot their strategies as needed. This agility allows organizations to stay ahead of the competition and seize new opportunities.

5. Strategic Alignment and Performance Optimization: EDI aligns decision-making processes with organizational strategies and goals. By providing a holistic view of the organization, EDI enables decision-makers to evaluate decisions in the context of financial performance, customer satisfaction, operational efficiency, and long-term sustainability. This alignment ensures that decisions are consistent with organizational priorities, optimizing overall performance and driving strategic growth.

6. Innovation and Competitive Advantage: EDI fosters a culture of innovation within organizations. By leveraging data analytics and generative models, EDI supports ideation, creativity, and exploration of new possibilities. It enables decision-makers to generate and evaluate innovative ideas, simulate scenarios, and assess potential outcomes. This focus on innovation gives organizations a competitive advantage by enabling them to identify emerging trends, disrupt markets, and deliver unique value to customers.

In conclusion, Enterprise Decision Intelligence has a significant impact on organizational performance and agility. By improving decision quality, enhancing operational efficiency, enabling proactive risk management, increasing agility, aligning strategies, and fostering innovation, EDI empowers organizations to make informed, data-driven decisions that drive performance and adaptability. As organizations continue to embrace EDI, they can gain a competitive edge in a rapidly evolving business landscape and achieve sustained success in their respective industries.

ı

Figure 11

Chapter 10: Applying Generative AI in Business Decision-Making

Generative models for predictive analysis: Forecasting, trend analysis, and demand prediction

Predictive analysis plays a crucial role in decision-making processes across various industries. Organizations strive to accurately forecast future trends, anticipate customer demand, and make informed decisions based on reliable predictions. In recent years, generative models have emerged as powerful tools for predictive analysis. These models, which include techniques such as generative adversarial networks (GANs) and variational autoencoders (VAEs), have the ability to generate new data that closely resembles existing patterns. In this section, we will explore how generative models can be used for predictive analysis, specifically focusing on forecasting, trend analysis, and demand prediction.

1. Forecasting:
Forecasting involves predicting future outcomes based on historical data patterns. Generative models can assist in forecasting by capturing complex patterns and dependencies in the data. By training a generative model on historical data, organizations can generate synthetic data that represents possible future scenarios. This synthetic data can then be used to project future trends and outcomes, aiding decision-making processes.

Generative models excel at capturing high-dimensional and nonlinear relationships, making them well-suited for forecasting tasks. They can generate multiple scenarios, allowing decision-makers to evaluate different potential futures and assess the associated risks and opportunities. By incorporating generative models into forecasting workflows, organizations can improve the accuracy and reliability of their predictions.

2. Trend Analysis:

Trend analysis involves identifying and understanding patterns in data over time. Generative models can aid in trend analysis by generating data points that align with existing trends. These models learn the underlying structure of the data and can generate new samples that follow the same patterns.

By using generative models for trend analysis, organizations can identify emerging trends, track the evolution of existing trends, and anticipate future developments. Decision-makers can gain insights into market dynamics, consumer preferences, and industry shifts. This knowledge enables organizations to make strategic decisions, adapt to changing trends, and stay ahead of the competition.

3. Demand Prediction:

Demand prediction is a critical aspect of supply chain management, sales forecasting, and inventory optimization. Generative models can be leveraged to predict customer demand by learning from historical sales data and generating synthetic demand patterns. These models can capture the complexities of consumer behavior and generate realistic demand scenarios.

By utilizing generative models for demand prediction, organizations can optimize inventory levels, allocate resources efficiently, and improve supply chain operations. Accurate demand predictions help minimize stockouts, reduce excess inventory, and enhance customer satisfaction. Generative models allow decision-makers to simulate different demand scenarios, assess the impact of marketing strategies, and optimize pricing and promotional activities.

Benefits and Considerations:

The use of generative models for predictive analysis offers several benefits. These models can capture complex patterns and dependencies that traditional statistical methods may miss. They provide flexibility in generating diverse scenarios and exploring alternative futures. Generative models can handle high-dimensional and nonlinear data, making them suitable for a wide range of prediction tasks. They also allow decision-makers to gain a deeper understanding of the underlying data dynamics.

However, it is important to consider some considerations when utilizing generative models for predictive analysis. The quality of the generated data depends on the quality and representativeness of the training data. Biases present in the training data can be reflected in the generated samples. Care must be taken to ensure that the training data is diverse, unbiased, and accurately represents the desired outcomes. Additionally, interpretability and explainability of generative models can be challenging, especially when making critical decisions based on their outputs.

In conclusion, generative models provide valuable capabilities for predictive analysis tasks such as forecasting, trend analysis, and demand prediction. By capturing complex patterns and generating synthetic data, these models enhance decision-making processes and improve the accuracy of predictions. Organizations can leverage generative models to gain insights, make informed decisions, and optimize operations in a dynamic and competitive business environment. As the field of generative modeling advances, the applications of these models in predictive analysis will continue to evolve, providing organizations with even more powerful tools for informed decision-making.

Optimization and Resource Allocation Using Generative AI

Optimizing resource allocation is a critical task for organizations across various industries. Efficiently allocating resources, such as manpower, budget, and equipment, can lead to cost savings, improved productivity, and enhanced overall performance. In recent years, the emergence of generative artificial intelligence (AI) techniques has provided new opportunities for organizations to tackle resource allocation challenges. In this section, we will explore how generative AI can be leveraged to optimize resource allocation.

Understanding Generative AI:

Generative AI refers to a class of AI techniques that focus on generating new content, such as images, text, or data, based on patterns and structures learned from existing examples. These models, including generative adversarial networks (GANs) and variational autoencoders (VAEs), have the ability to capture complex relationships and generate synthetic data that closely resembles the original data distribution.

Optimization in Resource Allocation:

Resource allocation optimization involves determining the most efficient way to distribute limited resources to meet specific objectives and constraints. Generative AI can play a vital role in resource allocation optimization by aiding decision-making processes. Here are some key ways in which generative AI can be applied:

1. Demand Prediction and Forecasting: Generative models can be trained on historical data to generate synthetic demand patterns. By accurately predicting future demand, organizations can optimize resource allocation based on anticipated needs. This can be particularly beneficial in supply chain management, production planning, and inventory optimization. Generative AI can provide insights into future demand scenarios, enabling organizations to allocate resources efficiently and reduce waste.

2. Scenario Analysis and Simulation: Generative models allow organizations to simulate different resource allocation scenarios and assess their impact. By generating synthetic data and exploring various possibilities, decision-makers can evaluate different resource allocation strategies and make informed decisions. This helps identify optimal allocation plans, manage risks, and assess trade-offs between competing objectives.

3. Portfolio Optimization: In financial and investment contexts, generative AI can assist in portfolio optimization. By generating synthetic asset returns based on historical data, generative models can enable decision-makers to evaluate and optimize investment portfolios. This helps identify the optimal allocation of financial resources, considering risk, return, and diversification.

4. Workforce and Shift Planning: Generative AI can aid in optimizing workforce allocation and shift planning. By analyzing historical data on work patterns, generative models can generate synthetic work schedules that meet demand while considering factors such as employee availability, skill sets, and work regulations. This helps organizations allocate human resources effectively, minimize overtime, and maintain productivity levels.

Benefits and Considerations:
Leveraging generative AI for resource allocation optimization offers several benefits. These include improved accuracy and efficiency in decision-making, enhanced utilization of resources, cost savings, and the ability to consider multiple scenarios. Generative models can capture complex relationships and generate diverse synthetic data, enabling decision-makers to explore a wide range of possibilities.

However, it is important to consider some considerations when using generative AI for resource allocation optimization. The quality of the generated data depends on the quality and representativeness of the training data. Biases present in the training data can be reflected in the generated samples, so careful attention must be paid to ensure fairness and avoid discrimination. Additionally, interpretability of generative models can be challenging, making it important to validate the generated results and understand the underlying assumptions and limitations.

In conclusion, generative AI provides organizations with powerful tools to optimize resource allocation. By leveraging generative models, organizations can improve demand prediction, conduct scenario analysis, optimize portfolios, and streamline workforce planning. The ability to generate synthetic data and explore various resource allocation strategies empowers decision-makers to make informed choices and maximize the efficiency and effectiveness of resource utilization. As generative AI continues to advance, its applications in resource allocation optimization will further contribute to improved organizational performance and decision-making capabilities.

Generative AI-Driven Customer Insights and Personalization: Unlocking the Power of Data

In the age of digital transformation and ever-increasing customer expectations, organizations are striving to deliver personalized experiences that resonate with their customers. Personalization is no longer a luxury but a necessity for businesses seeking to stay competitive and build long-term customer loyalty. One of the key enablers of personalized experiences is generative artificial intelligence (AI). Generative AI-driven customer insights harness the power of data and advanced algorithms to unlock valuable customer insights and deliver tailored experiences. In this section, we will explore how generative AI is transforming customer insights and personalization.

Understanding Generative AI-Driven Customer Insights:

Generative AI refers to a class of AI techniques that generate new content, such as images, text, or recommendations, based on patterns and structures learned from existing data. These techniques, including generative adversarial networks (GANs) and variational autoencoders (VAEs), have revolutionized the way organizations can understand and engage with their customers.

1. Customer Understanding and Segmentation: Generative AI allows organizations to analyze vast amounts of customer data to gain deeper insights and identify meaningful patterns. By training generative models on diverse customer data, organizations can uncover hidden segments and behavioral trends. This enables businesses to better understand their customers, their preferences, and their needs, allowing for more effective personalization strategies.

2. Enhanced Recommendation Systems: Generative AI-driven recommendation systems have transformed how businesses provide personalized recommendations to customers. These systems can generate accurate and relevant product recommendations based on a customer's past behavior, preferences, and similarities to other users. By leveraging generative AI, organizations can improve the accuracy and effectiveness of recommendation engines, leading to increased customer satisfaction and engagement.

3. Hyper-Personalized Experiences: Generative AI enables hyper-personalization by tailoring experiences to individual customers at a granular level. By analyzing customer data, including demographic information, purchase history, and browsing behavior, generative models can generate personalized content, offers, or product variations. This level of personalization helps organizations create unique and engaging experiences that resonate with individual customers, fostering loyalty and driving customer lifetime value.

Benefits of Generative AI-Driven Customer Insights and Personalization

1. Improved Customer Engagement: By leveraging generative AI-driven insights, organizations can deliver personalized experiences that captivate and engage customers. Personalized content, recommendations, and offers build stronger connections and enhance customer satisfaction. This leads to increased customer loyalty, higher conversion rates, and ultimately, business growth.

2. Enhanced Customer Retention: Personalization powered by generative AI-driven insights can significantly impact customer retention. When customers feel understood and catered to on an individual level, they are more likely to remain loyal to a brand. By continuously leveraging customer data and generative models, organizations can proactively anticipate customer needs, provide tailored solutions, and maintain strong customer relationships.

3. Increased Revenue and Upselling Opportunities: Personalization has a direct impact on revenue generation. By tailoring recommendations and offers to customer preferences, generative AI-driven personalization can drive upselling and cross-selling opportunities. When customers are presented with relevant and compelling recommendations, they are more inclined to make additional purchases, leading to increased revenue streams.

4. Data-Driven Decision-Making: Generative AI-driven customer insights provide organizations with a data-driven foundation for decision-making. By analyzing customer data at scale, generative models can extract valuable insights, uncover new market segments, and inform strategic decisions. This empowers organizations to make informed choices, optimize marketing campaigns, and allocate resources effectively.

Considerations for Generative AI-Driven Personalization

While the benefits of generative AI-driven customer insights and personalization are significant, organizations must consider ethical considerations and data privacy. It is crucial to ensure that customer data is handled responsibly, adhering to privacy regulations and obtaining appropriate consent. Additionally, organizations must be transparent about the use of generative AI and how customer data is utilized to provide personalized experiences.

In conclusion, generative AI-driven customer insights and personalization have transformed how organizations understand, engage, and serve their customers. By leveraging the power of data and advanced algorithms, businesses can unlock valuable insights, deliver personalized experiences, and cultivate long-lasting customer relationships. As generative AI continues to advance, organizations that embrace this technology and harness its capabilities will be better equipped to meet the evolving expectations of their customers in the digital era.

Enhancing risk assessment and management with Generative AI

Risk assessment and management are critical components of decision-making processes across industries. Organizations must navigate various risks, including financial uncertainties, operational challenges, regulatory compliance, and cybersecurity threats. To strengthen risk management practices, organizations are turning to generative artificial intelligence (AI) techniques. Generative AI, including generative adversarial networks (GANs) and variational autoencoders (VAEs), offers unique capabilities to enhance risk assessment and management. In this section, we will explore how generative AI can empower organizations to better understand and mitigate risks.

1. Identifying Hidden Patterns and Anomalies:
Generative AI can help organizations identify hidden patterns and anomalies in data, allowing for more accurate risk assessment. By training generative models on large datasets, organizations can gain insights into complex relationships and detect unusual patterns that might indicate potential risks. Generative AI techniques excel at capturing high-dimensional and non-linear dependencies, enabling the identification of subtle anomalies that may be challenging to detect with traditional methods. This enhanced understanding of hidden patterns and anomalies enhances risk assessment and allows for proactive risk mitigation.

2. Scenario Analysis and Risk Simulation:
Generative AI-driven models enable organizations to conduct scenario analysis and risk simulation. By generating synthetic data based on existing patterns, organizations can simulate various scenarios and assess the potential impact of different risk factors. This capability is particularly useful in complex and uncertain environments, where traditional risk assessment methods may fall short. Generative AI facilitates comprehensive risk analysis by providing decision-makers with insights into multiple plausible scenarios and their associated risks, enabling more effective risk management strategies.

3. Synthetic Data Generation for Risk Modeling:
Generative AI can assist in risk modeling by generating synthetic data that closely resembles real-world data distributions. This is particularly valuable in situations where obtaining sufficient

real-world data may be challenging or restricted due to privacy concerns. Generative models can learn from available data and generate synthetic data that preserves important statistical properties. This synthetic data can be used to build robust risk models, facilitating accurate risk assessments and improving decision-making processes.

4. Early Warning Systems and Risk Prediction:
Generative AI techniques can contribute to the development of early warning systems and risk prediction models. By analyzing historical data and capturing temporal dependencies, generative models can generate forecasts and predictions related to potential risks. Early identification of risks allows organizations to take proactive measures to prevent or mitigate the impact of adverse events. Generative AI-driven risk prediction models can provide decision-makers with timely and accurate information, facilitating proactive risk management and reducing the potential negative consequences of risks.

5. Optimization and Risk Mitigation:
Generative AI can support risk mitigation efforts by optimizing resource allocation and risk mitigation strategies. By leveraging generative models, organizations can simulate various resource allocation scenarios and assess their impact on risk mitigation. This allows decision-makers to identify optimal strategies that effectively balance risk mitigation and resource utilization. Generative AI-driven optimization models help organizations allocate resources efficiently, implement risk mitigation measures, and reduce exposure to potential risks.

Considerations for Generative AI in Risk Management

While generative AI offers significant benefits for risk assessment and management, several considerations should be taken into account:

1. Ethical Use of Data: Organizations must ensure that the use of generative AI techniques for risk management aligns with ethical guidelines and complies with data privacy regulations. Attention should be given to data privacy, data security, and the appropriate use of sensitive information.

2. Transparency and Interpretability: Generative AI models are often complex and may lack interpretability. Organizations should strive to understand and explain the outputs and decisions made by generative models to build trust and confidence in risk management practices.

3. Continuous Learning and Improvement: Generative AI models should be regularly updated and trained on up-to-date data to capture evolving risk patterns. Continuous learning and improvement ensure that risk assessments and predictions remain accurate and effective over time.

Conclusion:
Generative AI techniques provide organizations with valuable tools to enhance risk assessment and management. By identifying hidden patterns, conducting scenario analysis, generating synthetic data, enabling risk prediction, and optimizing risk mitigation strategies, generative AI empowers decision-makers to make informed choices and proactively address risks. Organizations that embrace generative AI in risk management gain a competitive advantage by improving risk assessment accuracy, increasing resilience, and facilitating more effective risk mitigation strategies.

Figure 12

Chapter 11: Building an Effective Generative AI Infrastructure

Data Governance and Management for Generative AI: Ensuring Ethical and Responsible Practices

Generative artificial intelligence (AI) techniques, such as generative adversarial networks (GANs) and variational autoencoders (VAEs), have transformed the way organizations generate new content based on existing patterns. However, the use of generative AI raises important considerations around data governance and management. To ensure ethical and responsible practices, organizations must prioritize data governance and management frameworks when leveraging generative AI. In this section, we will explore the significance of data governance and management in the context of generative AI.

1. Data Quality and Integrity:
Data governance and management are essential for maintaining data quality and integrity. Generative AI models heavily rely on the data they are trained on. Therefore, it is crucial to ensure that the training data is accurate, representative, and free from biases. Organizations need to establish data quality frameworks, implement data cleansing processes, and enforce data standards to minimize the risk of generating biased or misleading content. By maintaining high data quality standards, organizations can enhance the accuracy and reliability of generative AI outputs.

2. Privacy and Security:

Privacy and security considerations are paramount when working with data in generative AI applications. Organizations must handle sensitive data responsibly and comply with privacy regulations. Anonymization techniques should be applied to protect personally identifiable information (PII) when training generative AI models. Data encryption, access controls, and secure storage should be implemented to safeguard sensitive data from unauthorized access. By prioritizing privacy and security measures, organizations can ensure that generative AI applications do not compromise data confidentiality or violate privacy rights.

3. Consent and Transparency:

Organizations must ensure transparency and seek appropriate consent when using data for generative AI applications. Clear communication about the purpose, use, and potential impacts of generative AI-generated content is crucial. Transparent disclosure helps build trust with individuals whose data is being used and allows them to make informed decisions about their data. Organizations should provide clear guidelines and obtain explicit consent when collecting, processing, and utilizing data for generative AI applications. Transparency and consent form the foundation of ethical data practices in the context of generative AI.

4. Bias and Fairness:

Generative AI models can inadvertently inherit biases present in the training data. It is essential to proactively identify and mitigate biases to ensure fairness in generative AI outputs. Organizations should implement strategies to assess and address biases, conduct regular audits of generative AI models, and invest in diverse and representative training data. By actively mitigating biases, organizations can minimize the risk of generating discriminatory content and promote fairness in generative AI-generated outputs.

5. Model Interpretability and Explainability:

Generative AI models are often complex and may lack interpretability. Ensuring model interpretability and explainability is crucial for building trust, understanding how generative AI models work, and identifying potential biases or errors. Organizations should strive to develop methods to interpret and explain generative AI-generated content. By promoting model interpretability and explainability, organizations can provide insights into how generative AI models generate content and increase transparency in decision-making processes.

6. Governance Frameworks:

Establishing robust data governance frameworks is crucial for responsible generative AI usage. Organizations should define clear policies, guidelines, and procedures for data collection, storage, processing, and sharing. Roles and responsibilities should be defined to ensure accountability and compliance with data governance practices. Regular audits and reviews of generative AI systems should be conducted to identify potential risks, ensure compliance, and address emerging challenges. A comprehensive governance framework sets the foundation for ethical and responsible generative AI practices.

In conclusion, data governance and management are integral to responsible and ethical generative AI usage. By focusing on data quality, privacy, transparency, bias mitigation, interpretability, and governance frameworks, organizations can ensure the ethical and responsible use of generative AI. Adhering to robust data governance and management practices empowers organizations to leverage generative AI techniques while upholding ethical standards, building trust, and mitigating risks associated with generative AI applications.

Developing robust generative models: Techniques, algorithms, and model architectures.

Generative models, such as generative adversarial networks (GANs) and variational autoencoders (VAEs), have revolutionized the field of artificial intelligence (AI) by enabling the creation of new content, including images, text, and music. Developing robust generative models requires a combination of techniques, algorithms, and model architectures to ensure high-quality outputs and address challenges such as mode collapse and lack of diversity. In this section, we will explore the key considerations and approaches involved in developing robust generative models.

1. GANs:

Generative adversarial networks (GANs) are a popular class of generative models that consist of a generator and a discriminator. The generator generates synthetic data, while the discriminator tries to distinguish between real and synthetic data. Training GANs involves an adversarial process where the generator and discriminator compete against each other, driving the generator to produce more realistic outputs.

To develop robust GANs, several techniques can be employed:

- Architectural Improvements: Architectural modifications, such as using deep convolutional layers in computer vision tasks or recurrent neural networks in text generation, can improve the performance and quality of GANs.
- Regularization Techniques: Regularization techniques, such as adding dropout layers or using batch normalization, can help prevent overfitting and improve the generalization of GAN models.
- Training Stabilization: GAN training can be unstable, leading to mode collapse or poor convergence. Techniques such as Wasserstein GAN (WGAN) and spectral normalization can stabilize training and improve the performance of GAN models.

2. VAEs:

Variational autoencoders (VAEs) are another widely used generative model framework. VAEs consist of an encoder network that maps input data to a latent space and a decoder network that reconstructs the input data from the latent space. VAEs learn a latent representation of the input data distribution and generate new samples by sampling from the learned latent space.

To develop robust VAEs, the following techniques can be employed:
- Latent Space Regularization: Techniques like regularization methods (e.g., variational dropout) or adversarial regularizers can help improve the diversity and quality of the generated samples.
- Improved Latent Space Representation: Exploring more expressive latent space structures, such as disentangled representations or hierarchical structures, can enable better control and generation of diverse samples.
- Reconstruction Loss Optimization: Optimizing the reconstruction loss can help ensure that the generated samples closely resemble the input data, leading to higher quality generative models.

3. Evaluation Metrics:

Developing robust generative models requires effective evaluation metrics to measure their performance. Common metrics include:
- Inception Score: It measures the quality and diversity of generated samples based on the

predictions of an Inception classifier.
- Frechet Inception Distance (FID): It calculates the distance between the distribution of real data and generated samples in feature space.
- Perceptual Metrics: Metrics such as LPIPS (Learned Perceptual Image Patch Similarity) or MS-SSIM (Multi-Scale Structural Similarity) assess the perceptual quality of generated samples.

It is important to use multiple evaluation metrics to gain a comprehensive understanding of the generative model's performance.

4. Data Augmentation and Preprocessing:

Data augmentation and preprocessing techniques play a crucial role in developing robust generative models. Techniques such as rotation, translation, scaling, and noise injection can augment the training data and improve the diversity of generated samples. Preprocessing steps like normalization, denoising, or dimensionality reduction can also enhance the quality of generative models.

5. Adapting Loss Functions:

Tailoring loss functions to the specific task and characteristics of the data can improve the performance and stability of generative models. For example, using Wasserstein loss instead of traditional adversarial loss in GANs can lead to better training stability and sample quality.

6. Transfer Learning and Pretrained Models:

Leveraging transfer learning and pretrained models can expedite the development of robust generative models. By leveraging models pretrained on large-scale datasets, such as ImageNet, and fine-tuning them on the target dataset, one can benefit from the learned representations and accelerate the training process.

In conclusion, developing robust generative models involves a combination of techniques, algorithms, and model architectures. It requires careful consideration of architectural improvements, regularization techniques, training stabilization, and evaluation metrics. Data augmentation, preprocessing, loss function adaptation, and the use of transfer learning can further enhance the robustness and performance of generative models. As the field of generative modeling continues to advance, these techniques will play a vital role in creating high-quality generative models with diverse applications in various domains.

Integrating Generative AI into Existing Decision-Making Frameworks: Harnessing the Power of Synthetic Data and Enhanced Insights

As organizations strive to make informed decisions in a rapidly evolving business landscape, the integration of artificial intelligence (AI) technologies has become increasingly vital. Generative AI, in particular, offers unique capabilities to generate synthetic data and provide enhanced insights. By integrating generative AI into existing decision-making frameworks, organizations can leverage the power of synthetic data and augment their decision-making processes. In this section, we will explore the benefits and considerations of integrating generative AI into existing decision-making frameworks.

1. Generating Synthetic Data:

Generative AI techniques, such as generative adversarial networks (GANs) and variational autoencoders (VAEs), can generate synthetic data that closely resembles real-world data patterns. By integrating generative AI, organizations can expand their data pool and address data scarcity issues. This synthetic data can be used to augment existing datasets, enrich feature sets, and provide additional perspectives for decision-making. Incorporating synthetic data enables decision-makers to gain insights into potential scenarios, explore alternative possibilities, and make more informed choices.

2. Augmenting Decision-Making Insights:

Generative AI models can uncover hidden patterns, relationships, and trends in data. By integrating generative AI techniques, organizations can enhance their decision-making insights. These models can generate new samples and alternative scenarios that decision-makers may not have considered. This expands the decision space, enables the exploration of different options, and facilitates better evaluation of risks and opportunities. The augmented insights provided by generative AI empower decision-makers with a broader and more comprehensive view of the decision landscape.

3. Addressing Data Imbalance and Bias:

Data imbalance and bias are common challenges in decision-making processes. Integrating generative AI can help address these issues by generating synthetic data that balances representation across different classes or demographics. Generative models can learn from the available data and produce synthetic samples for that ensure fairness and reduce bias in decision-making. This integration enables decision-makers to account for a wider range of perspectives and avoid potential biases that may exist in the original dataset.

4. Simulating Scenarios and What-If Analysis:

Generative AI models allow for scenario simulation and what-if analysis, empowering decision-makers to assess the potential outcomes of different choices. By generating synthetic data that represents alternative scenarios, organizations can explore the impact of various decisions before implementation. This capability aids in risk assessment, resource allocation, and strategic planning. Decision-makers can identify potential challenges, evaluate trade-offs, and make more informed decisions by simulating scenarios with generative AI.

Considerations for Integration

While integrating generative AI into existing decision-making frameworks offers significant advantages, organizations must consider several factors:

1. Data Privacy and Security: Organizations must ensure that privacy and security protocols are in place when integrating generative AI. Handling sensitive data requires compliance with relevant regulations and ethical considerations. Data anonymization, secure storage, and appropriate access controls are crucial to protect privacy and maintain data security.

2. Model Interpretability and Explainability: Generative AI models can be complex, making it challenging to interpret their outputs and understand the decision-making process. Organizations should strive for model interpretability and explainability to build trust and confidence. Ensuring that generative AI models are transparent and explainable helps decision-makers understand how the models generate insights and supports their interpretation.

3. Ethical Considerations and Bias Mitigation: Generative AI models can inadvertently amplify biases present in the training data. Organizations must be vigilant in identifying and mitigating biases to ensure fair and ethical decision-making. Regular monitoring, validation, and bias mitigation techniques should be applied to generative AI models to minimize the risk of perpetuating or exacerbating biases in decision-making.

4. Collaboration and Change Management: Integrating generative AI into existing decision-making frameworks requires collaboration among various stakeholders. Effective change management practices should be implemented to facilitate the integration process and ensure a smooth transition. Education and training on generative AI, its benefits, and its implications are essential for decision-makers and other personnel involved in the process.

In conclusion, integrating generative AI into existing decision-making frameworks offers organizations the opportunity to harness the power of synthetic data and enhanced insights. By generating synthetic data, augmenting decision-making insights, addressing data imbalances and biases, and enabling scenario simulation, generative AI empowers decision-makers to make more informed choices. However, careful attention must be given to data privacy, model interpretability, ethical considerations, and change management to ensure successful integration and responsible use of generative AI in decision-making processes.

Balancing Interpretability and Complexity in Generative AI Models: Striking the Right Trade-Off

Generative artificial intelligence (AI) models, such as generative adversarial networks (GANs) and variational autoencoders (VAEs), have demonstrated remarkable capabilities in generating new content based on learned patterns. However, as these models grow in complexity and power, a key challenge arises: balancing interpretability with the inherent complexity of generative AI models. In this section, we will explore the importance of striking the right trade-off between interpretability and complexity in generative AI models.

Interpretability refers to the ability to understand and explain how a model arrives at its decisions or generates specific outputs. Complexity, on the other hand, refers to the intricacy and sophistication of the model architecture and its underlying algorithms. Striking a balance between these two aspects is crucial for several reasons:

1. Trust and Transparency:

Interpretability plays a vital role in building trust and transparency with stakeholders. Understanding how a generative AI model operates and generates outputs fosters confidence in its reliability and fairness. Decision-makers and end-users need to have a clear understanding of the underlying mechanisms to trust the model's predictions or generated content. Striking the right balance between interpretability and complexity enables the model to be more transparent and comprehensible to users, thereby enhancing trust in its outcomes.

2. Ethical Considerations:

As generative AI models become more powerful, ethical considerations come to the forefront. It is essential to identify and address any biases or unfairness that may be present in the model's outputs.

Interpretable models provide insights into the decision-making process, allowing for the detection and mitigation of biases. Balancing interpretability and complexity facilitates ethical decision-making and enables organizations to uphold fairness and accountability.

3. Regulatory Compliance:

Many industries are subject to regulatory frameworks that mandate transparency and explainability in AI-driven systems. Striking a balance between interpretability and complexity ensures compliance with regulatory requirements. Models that can be audited and explained demonstrate accountability and facilitate compliance with legal and regulatory obligations. Organizations can avoid potential legal complications and ensure a responsible deployment of generative AI models by considering interpretability alongside complexity.

4. User Adoption and Acceptance:

Complex models might be intimidating or difficult for users to comprehend, hindering their adoption and acceptance. Users are more likely to embrace and utilize generative AI models when they can understand and interpret the model's outputs. Simplicity and interpretability make the models more accessible and user-friendly, enabling broader adoption across various domains and user groups.

Strategies for Balancing Interpretability and Complexity

1. Simplifying Model Architectures:

One approach to balancing interpretability and complexity is to simplify the model architecture. This involves reducing the number of layers, parameters, or components in the model to make it more comprehensible. Simpler models are often easier to interpret, allowing decision-makers and users to gain insights into the model's decision-making process.

2. Incorporating Explainable AI Techniques:

Explainable AI techniques can be applied to complex generative models to improve interpretability. Methods like attention mechanisms, feature visualization, or model-agnostic techniques (e.g., LIME or SHAP) help highlight important features and provide explanations for the model's outputs. These techniques enhance interpretability without sacrificing the model's complexity.

3. Providing Post-hoc Explanations:

Post-hoc explanation methods focus on generating explanations for the model's outputs after they are produced. Techniques such as counterfactual explanations or rule-based approaches can help explain the reasoning behind the model's decisions. These explanations can be provided alongside the model's outputs to enhance interpretability.

4. Collaborating with Domain Experts:

Collaboration between AI practitioners and domain experts can greatly aid in achieving a balance between interpretability and complexity. Domain experts can provide insights into the specific requirements and expectations of the application domain, ensuring that the model's outputs are meaningful and interpretable within the context of the problem at hand.

In conclusion, finding the right balance between interpretability and complexity in generative AI models is crucial for building trust, addressing ethical considerations, ensuring regulatory compliance, and facilitating user adoption. By simplifying model architectures, incorporating explainable AI techniques, providing post-hoc explanations, and collaborating with domain experts, organizations can strike an optimal trade-off that allows for meaningful interpretation while leveraging the power of complex generative AI models. A thoughtful approach to balancing interpretability and complexity paves the way for responsible and impactful use of generative AI in various domains and applications.

Ensuring Ethical Considerations and Regulatory Compliance in Enterprise Decision Intelligence: A Responsible Approach

Enterprise Decision Intelligence (EDI) has emerged as a critical discipline that combines data, analytics, and artificial intelligence (AI) to inform and optimize decision-making processes in organizations. While EDI holds tremendous potential for driving business growth and innovation, it must be approached with a strong focus on ethical considerations and regulatory compliance. In this section, we will explore the importance of ensuring ethical practices and regulatory compliance in EDI and discuss key considerations for organizations.

1. Data Privacy and Consent:
Data privacy is a fundamental aspect of ethical EDI. Organizations must handle data in accordance with relevant privacy laws and regulations, such as the General Data Protection Regulation (GDPR) or the California Consumer Privacy Act (CCPA). It is crucial to obtain proper consent for data collection, storage, and processing, ensuring transparency about how data will be used in EDI initiatives. Anonymization techniques, encryption, and secure data storage are essential for safeguarding sensitive information and protecting individuals' privacy rights.

2. Fairness and Bias Mitigation:
EDI should be designed and implemented in a way that ensures fairness and mitigates biases. Algorithms and models used in EDI should be regularly monitored and evaluated to identify any biases that may emerge during data collection, preprocessing, or model training. Mitigation techniques, such as ensuring diverse and representative training data, conducting bias audits, and employing fairness-aware algorithms, can help minimize the risk of biased decision-making and discriminatory outcomes.

3. Transparency and Explainability:
Transparency and explainability are critical for building trust and accountability in EDI. Organizations should strive to make the decision-making process transparent and provide explanations for the outcomes produced by AI models. Techniques such as model interpretability, feature importance analysis, and post-hoc explanation methods can help shed light on the reasoning behind decisions made by EDI systems. Transparency enables stakeholders, including decision-makers, regulators, and customers, to understand and evaluate the fairness, reliability, and ethical implications of EDI-generated insights.

4. Regulatory Compliance:
EDI initiatives must adhere to relevant laws, regulations, and industry-specific guidelines. Organizations should stay abreast of legal and regulatory requirements, such as those related to data protection, consumer rights, and industry-specific compliance standards. Compliance with regulations such as GDPR, CCPA, HIPAA (Health Insurance Portability and Accountability Act),

or financial industry regulations ensures that EDI initiatives align with legal obligations and best practices in data governance and management.

5. Human Oversight and Accountability:

While EDI leverages AI and automation, human oversight and accountability remain crucial. Organizations should establish clear roles and responsibilities for decision-making, ensuring that humans are ultimately accountable for the outcomes of EDI initiatives. Humans play a critical role in defining the objectives, setting ethical guidelines, monitoring the performance of AI models, and making ethical judgments when complex or nuanced decisions arise. Regular audits, ethical training, and robust governance frameworks help maintain human accountability in EDI processes.

6. Continuous Monitoring and Evaluation:

Ethical considerations and regulatory compliance in EDI require ongoing monitoring and evaluation. Organizations should regularly assess the impact of EDI systems on fairness, privacy, and compliance. This includes evaluating the performance and biases of AI models, monitoring data usage and consent management practices, and conducting risk assessments to identify and address potential ethical and regulatory concerns. Continuous monitoring and evaluation ensure that EDI initiatives remain aligned with ethical principles and regulatory requirements as technology and business contexts evolve.

In conclusion, ensuring ethical considerations and regulatory compliance in EDI is essential for maintaining trust, fairness, and accountability. Organizations must prioritize data privacy, fairness, transparency, and compliance with relevant regulations. By integrating ethical practices and compliance measures into EDI initiatives, organizations can harness the power of data and AI while upholding ethical standards and meeting legal obligations. A responsible approach to EDI not only benefits organizations but also fosters trust among stakeholders and contributes to sustainable and responsible business practices.

"Hard work, determination, success!"

Figure 13

Chapter 12: Overcoming Challenges and Maximizing the Potential

Addressing Bias and Fairness in Generative AI Models:
Striving for Ethical and Inclusive Outputs

Generative artificial intelligence (AI) models, such as generative adversarial networks (GANs) and variational autoencoders (VAEs), have the power to generate new content based on learned patterns. However, the outputs of these models are not immune to biases that may exist in the training data or emerge during the model's learning process. Addressing bias and promoting fairness in generative AI models is crucial for ensuring ethical and inclusive outcomes. In this section, we will explore the importance of addressing bias and fairness in generative AI models and discuss key strategies for achieving this goal.

Understanding Bias in Generative AI Models:
Bias in generative AI models refers to the systematic favoritism or discrimination in the generated content. This bias can arise from various sources, including biases present in the training data, biased data collection processes, or limitations in the model architecture and algorithms. Addressing bias is essential to avoid perpetuating stereotypes, unfair representation, or discriminatory outputs that can have negative societal impacts.

Strategies for Addressing Bias and Promoting Fairness

1. Diverse and Representative Training Data:
The quality and representativeness of training data are critical factors in addressing bias. To promote fairness, it is essential to ensure that the training data encompasses diverse perspectives and adequately represents the target population. Careful consideration should be given to including data from different demographics, socioeconomic backgrounds, cultures, and geographic locations. Additionally, data collection processes should be designed to minimize selection biases and capture a broad range of experiences.

2. Preprocessing and Data Cleaning:
Before training generative AI models, it is crucial to carefully preprocess and clean the data to remove any inherent biases. This includes identifying and addressing potential biases in labels, annotations, or metadata associated with the training data. Data cleaning techniques such as debiasing algorithms, fairness-aware preprocessing, or outlier detection can help reduce biases and ensure a more balanced representation in the training data.

3. Bias Detection and Mitigation:
Implementing bias detection and mitigation techniques during the training and evaluation phases is essential. Bias detection methods can help identify potential biases in the model outputs by analyzing the generated content across different demographic groups or protected attributes. If biases are detected, mitigation strategies, such as modifying loss functions, introducing fairness constraints, or using adversarial training, can be employed to reduce bias and promote fairness.

4. Regular Auditing and Evaluation:
Ongoing auditing and evaluation of generative AI models are crucial to monitor and mitigate biases. Regular assessments should be conducted to identify any biases that may have emerged during model training or deployment. These audits can involve analyzing model performance across different demographic groups, evaluating fairness metrics, and soliciting feedback from diverse stakeholders to understand the impact of the generated content.

5. User Feedback and Iterative Improvement:
User feedback plays a vital role in addressing bias and promoting fairness. Organizations should actively seek input from users and stakeholders to gain insights into potential biases or unintended consequences of the generated content. This feedback can inform iterative improvements to the generative AI models, ensuring that biases are identified, rectified, and continuously monitored throughout the development and deployment cycles.

6. Interdisciplinary Collaboration:
Addressing bias and promoting fairness requires collaboration among diverse stakeholders, including data scientists, domain experts, ethicists, and individuals from affected communities. Interdisciplinary collaboration helps identify potential biases, design effective mitigation strategies, and ensure that generative AI models are developed in a manner that aligns with ethical standards and societal values.

In conclusion, addressing bias and promoting fairness in generative AI models is a vital step toward responsible and ethical AI deployment. By employing strategies such as diverse training data, preprocessing, bias detection and mitigation, regular auditing, user feedback, and interdisciplinary collaboration, organizations can mitigate biases, promote fairness, and ensure that generative AI models contribute to inclusive and equitable outcomes. Striving for bias-free and fair generative AI models will foster trust, respect diversity, and facilitate the development of AI systems that truly benefit all stakeholders.

Privacy, Security, and Data Protection in Generative AI Applications: Safeguarding Confidentiality and Ethical Use

Generative artificial intelligence (AI) applications, such as generative adversarial networks (GANs) and variational autoencoders (VAEs), have revolutionized the way organizations generate new content based on existing patterns. However, the utilization of generative AI raises important considerations surrounding privacy, security, and data protection. Safeguarding the confidentiality of data and ensuring ethical use are crucial aspects of responsible generative AI deployment. In this section, we will delve into the significance of privacy, security, and data protection in generative AI applications and explore key considerations for organizations.

1. Data Privacy and Confidentiality:
Data privacy is a fundamental principle that organizations must adhere to when leveraging generative AI applications. It is crucial to handle data in accordance with relevant privacy regulations, such as the General Data Protection Regulation (GDPR) or the California Consumer Privacy Act (CCPA). Organizations must ensure that personally identifiable information (PII) is protected and processed only with appropriate consent. Anonymization techniques, data aggregation, and encryption methods should be employed to preserve privacy and confidentiality.

2. Secure Data Storage and Transfer:
Generative AI applications require access to large volumes of data, which necessitates secure data storage and transfer mechanisms. Organizations should implement robust security measures to protect data from unauthorized access, breaches, or misuse. This includes encryption of data at rest and in transit, access controls, secure data centers, and regular security audits. By ensuring the confidentiality and integrity of data, organizations can mitigate the risks associated with data breaches and unauthorized access.

3. Ethical Use and Compliance:
Ethical considerations play a critical role in the development and deployment of generative AI applications. Organizations should adhere to ethical guidelines and ensure that generative AI models are used for legitimate and responsible purposes. Transparency and clear communication about the use of generative AI, including data collection, storage, and processing, foster trust and compliance with ethical standards. Compliance with legal and regulatory frameworks, such as GDPR or industry-specific guidelines, is imperative to ensure ethical data practices and responsible use of generative AI.

4. Data Governance and Access Controls:
A robust data governance framework is essential to safeguard privacy, security, and data protection in generative AI applications. Organizations should establish clear policies, procedures, and guidelines for data collection, usage, retention, and sharing. Data access controls should be implemented to ensure that only authorized individuals have access to sensitive data. Regular audits and monitoring mechanisms should be in place to track data usage and ensure compliance with

established policies.

5. Minimizing Data Exposure and Retention:
To mitigate privacy risks, organizations should minimize data exposure and retention. Only the necessary data required for generative AI applications should be collected and stored. Data minimization principles ensure that organizations only retain data for as long as it is necessary and delete or anonymize data that is no longer needed. By adopting a minimalistic approach to data collection and retention, organizations can reduce the potential risks associated with data breaches and privacy violations.

6. Transparent Model Behavior and Explainability:
Generative AI models should exhibit transparent behavior and explainability to ensure ethical use and compliance. Organizations should strive to understand and explain the decision-making process of generative AI models, including the generation of content. Techniques such as interpretability methods, attention mechanisms, or model-agnostic approaches can help shed light on the model's behavior and enhance transparency. Clear explanations of model outputs and decisions build trust and allow for the identification of potential biases or ethical concerns.

In conclusion, privacy, security, and data protection are paramount in generative AI applications. Organizations must prioritize data privacy, ensure secure storage and transfer, and adhere to ethical guidelines and regulatory compliance. By implementing robust data governance practices, minimizing data exposure and retention, and promoting transparency and explainability, organizations can harness the power of generative AI while upholding privacy rights, maintaining data security, and adhering to ethical standards. Responsible and ethical deployment of generative AI applications ensures the trust of individuals, protects sensitive information, and fosters a positive impact on society.

Managing Organizational Change and Fostering a Culture of Decision Intelligence: Building a Foundation for Success

In today's fast-paced and data-driven business environment, organizations must continuously evolve and adapt to remain competitive. Embracing decision intelligence, which combines data, analytics, and AI to inform decision-making, is a transformative approach that empowers organizations to make more informed and effective choices. However, successfully implementing decision intelligence requires more than just technological integration. It demands managing organizational change and fostering a culture that embraces data-driven decision-making. In this section, we will explore strategies for managing organizational change and cultivating a culture of decision intelligence.

1. Leadership Support and Alignment:
Successful organizational change begins with leadership support and alignment. Executives and senior leaders should champion the adoption of decision intelligence and articulate a clear vision for its implementation. They need to communicate the benefits of data-driven decision-making, set expectations, and allocate resources to support the transformation. Leadership commitment helps create a sense of urgency and establishes a strong foundation for driving change throughout the organization.

2. Communication and Stakeholder Engagement:

Effective communication and stakeholder engagement are crucial for managing change and fostering a culture of decision intelligence. Regular and transparent communication about the goals, benefits, and progress of decision intelligence initiatives helps build buy-in and mitigate resistance. Engaging stakeholders from all levels of the organization, including employees, managers, and teams, fosters a sense of ownership and encourages active participation in the transformation process.

3. Training and Upskilling:

To build a culture of decision intelligence, organizations need to invest in training and upskilling employees. This includes providing comprehensive training programs on data literacy, analytics, and AI tools and techniques. Employees should be equipped with the necessary skills to understand and interpret data, make data-driven decisions, and leverage decision intelligence tools effectively. Ongoing learning opportunities and access to resources support the development of a data-driven mindset across the organization.

4. Empowering Decision-Makers:

A culture of decision intelligence thrives when decision-makers are empowered to utilize data and insights in their decision-making processes. Organizations should foster an environment where employees feel comfortable challenging assumptions, experimenting with data, and incorporating insights into their decisions. This requires providing access to relevant data, analytics platforms, and decision support tools. Empowered decision-makers are more likely to embrace data-driven approaches and contribute to a culture of continuous improvement.

5. Collaboration and Cross-Functional Teams:

Decision intelligence is most effective when it is embraced as a collaborative effort across different functions and teams. Encouraging cross-functional collaboration fosters the exchange of knowledge, expertise, and diverse perspectives. By bringing together individuals from various disciplines, such as data science, business operations, and strategy, organizations can leverage a wide range of insights to make more holistic and informed decisions.

6. Celebrating Success and Recognizing Data-Driven Efforts:

Recognizing and celebrating successes in data-driven decision-making reinforces the culture of decision intelligence. Organizations should acknowledge and reward individuals and teams that embrace data-driven practices and achieve positive outcomes through their decision-making efforts. This recognition not only motivates employees but also reinforces the value of data-driven approaches, fostering a culture where data is seen as a strategic asset.

7. Continuous Improvement and Learning:

Cultivating a culture of decision intelligence is an ongoing process that requires continuous improvement and learning. Organizations should encourage a mindset of curiosity, experimentation, and learning from failures. Regularly evaluating the impact of decision intelligence initiatives, soliciting feedback from stakeholders, and adapting strategies based on lessons learned are essential for continuous growth and improvement.

In conclusion, managing organizational change and fostering a culture of decision intelligence requires a comprehensive approach that encompasses leadership support, effective communication, training, empowerment, collaboration, and a focus on continuous improvement. By embracing these strategies, organizations can establish a strong foundation for data-driven decision-making, enhance their competitive advantage, and drive success in today's dynamic business landscape.

Overcoming Technical and Implementation Challenges: Navigating the Path to Successful Adoption

Implementing new technologies and approaches within an organization often comes with its fair share of technical and implementation challenges. The adoption of decision intelligence, data analytics, and artificial intelligence (AI) is no exception. Organizations must navigate these challenges effectively to ensure a smooth and successful implementation. In this section, we will explore common technical and implementation challenges faced during the adoption of decision intelligence and discuss strategies to overcome them.

1. Data Quality and Accessibility:

One of the primary challenges in implementing decision intelligence is ensuring the quality and accessibility of data. Poor data quality, such as missing values, inconsistencies, or inaccuracies, can hinder the effectiveness of decision-making processes. To overcome this challenge, organizations should establish data governance practices that include data cleaning, standardization, and validation processes. Investing in data infrastructure and tools that enable efficient data collection, storage, and retrieval can also enhance data accessibility and quality.

2. Integration and Compatibility:

Integrating decision intelligence systems with existing technologies and systems can be complex. Organizations may face challenges in compatibility, interoperability, or data exchange between different platforms. To address this, it is crucial to conduct a thorough analysis of existing systems, identify integration requirements, and leverage technologies that support seamless integration. Application programming interfaces (APIs), data connectors, and middleware solutions can facilitate the smooth exchange of data between systems.

3. Technical Expertise and Talent:

Implementing decision intelligence requires individuals with the necessary technical expertise and skills. Organizations may face challenges in finding and retaining talent with proficiency in data analytics, AI, and decision intelligence methodologies. To overcome this, organizations should invest in training and upskilling programs to develop internal talent. Collaboration with external consultants, partnerships with educational institutions, or hiring experienced professionals can also address the skills gap.

4. Change Management and User Adoption:

Resistance to change and lack of user adoption can pose significant challenges during the implementation of decision intelligence. Users may be accustomed to existing decision-making processes or skeptical about the benefits of data-driven approaches. Effective change management strategies should be employed to communicate the benefits, provide training and support, and address concerns or resistance. Involving users throughout the implementation process, soliciting their feedback, and demonstrating tangible results can foster a culture of acceptance and enthusiasm for the new approach.

5. Scalability and Infrastructure:

As organizations scale their decision intelligence initiatives, they may encounter challenges related to infrastructure and scalability. Increased data volume, computational requirements, and storage needs may strain existing infrastructure. It is essential to plan for scalability from the outset, considering factors such as cloud-based solutions, elastic computing resources, and distributed data storage. Collaboration with IT teams and infrastructure planning can help address scalability challenges effectively.

6. Ethical and Legal Considerations:

Implementing decision intelligence requires careful consideration of ethical and legal aspects. Organizations must navigate issues such as data privacy, security, compliance with regulations (e.g., GDPR, CCPA), and responsible AI usage. To overcome these challenges, organizations should establish robust data governance frameworks, prioritize privacy and security measures, and engage legal experts to ensure compliance with applicable laws and regulations. Regular audits and evaluations should be conducted to address ethical concerns and monitor the impact of decision intelligence on stakeholders.

7. Iterative Approach and Continuous Improvement:

Implementing decision intelligence is an ongoing process of learning, adapting, and improving. It is crucial to adopt an iterative approach that allows for experimentation, feedback incorporation, and continuous improvement. Regularly reviewing and assessing the performance of decision intelligence systems, monitoring outcomes, and soliciting user feedback enable organizations to identify and address technical or implementation challenges proactively.

In conclusion, overcoming technical and implementation challenges is vital for successful adoption of decision intelligence within organizations. By prioritizing data quality, ensuring compatibility, fostering technical expertise, managing change effectively, planning for scalability, addressing ethical considerations, and adopting an iterative approach, organizations can navigate these challenges and unlock the full potential of decision intelligence for informed and effective decision-making.

Scaling and Operationalizing Generative AI Solutions:

Unlocking Value and Efficiency

Generative artificial intelligence (AI) solutions, such as generative adversarial networks (GANs) and variational autoencoders (VAEs), have demonstrated their ability to generate new content and insights. However, successfully scaling and operationalizing generative AI solutions can be a complex endeavor. It requires careful planning, infrastructure considerations, and effective management to unlock the full value and efficiency of these solutions. In this section, we will explore strategies for scaling and operationalizing generative AI solutions to drive business growth and innovation.

1. Define Clear Objectives and Use Cases:

Before scaling generative AI solutions, it is essential to define clear objectives and identify relevant use cases. Understand the specific business problems or opportunities that generative AI can address. Whether it is generating synthetic data, creating personalized recommendations, or enhancing creative content, aligning the objectives with the organization's strategic goals ensures that scaling efforts are focused and impactful.

2. Infrastructure and Resource Planning:

Scaling generative AI solutions requires careful infrastructure and resource planning. Consider the computational requirements, storage capacity, and networking capabilities necessary to handle the increased workload. Cloud-based platforms, elastic computing resources, and scalable storage solutions can provide the flexibility and scalability needed to support generative AI at scale. Assessing the infrastructure needs and planning resources accordingly help ensure smooth operations as the solution expands.

3. Data Management and Governance:

Effective data management and governance are critical for scaling generative AI solutions. Establish robust data governance practices to ensure data quality, integrity, and security. Implement data pipelines and workflows that can handle large volumes of data efficiently. Consider data anonymization techniques, privacy measures, and compliance with relevant regulations to protect sensitive information. Efficient data management allows for smooth scaling and enhances the reliability and trustworthiness of generative AI outputs.

4. Model Optimization and Efficiency:

As generative AI solutions scale, optimizing models and improving efficiency becomes paramount. Consider techniques like model compression, parameter tuning, and architecture enhancements to improve computational efficiency and reduce resource requirements. Continuously monitor and evaluate model performance to identify bottlenecks or areas for improvement. Iterative optimization ensures that generative AI solutions remain efficient and cost-effective as they scale.

5. Automation and DevOps:

Automation and DevOps practices streamline the deployment, monitoring, and maintenance of generative AI solutions. Adopt continuous integration and continuous deployment (CI/CD) pipelines to automate the deployment process and enable rapid iterations. Implement robust monitoring and alerting mechanisms to track performance, detect anomalies, and address issues proactively. DevOps practices enable efficient management of generative AI solutions and support scalability.

6. Collaboration and Knowledge Sharing:

Scaling generative AI solutions requires collaboration and knowledge sharing across teams and departments. Foster a culture of collaboration by establishing cross-functional teams and encouraging knowledge exchange. Share best practices, lessons learned, and insights gained from scaling efforts. Encourage data scientists, engineers, and domain experts to collaborate closely, leveraging their collective expertise to drive successful scaling and operationalization.

7. Continuous Learning and Improvement:

Generative AI is an evolving field, and continuous learning is essential for successful scaling and operationalization. Stay updated with the latest research, techniques, and advancements in generative AI. Foster a learning culture that encourages experimentation, encourages feedback, and incorporates new learnings into the generative AI solution. Regularly evaluate the impact of generative AI outputs and solicit user feedback to drive continuous improvement and enhance the value delivered by the solution.

In conclusion, scaling and operationalizing generative AI solutions require strategic planning, infrastructure considerations, efficient data management, and collaboration. By defining clear objectives, planning infrastructure and resources, implementing effective data governance, optimizing models, embracing automation and DevOps practices, fostering collaboration, and promoting continuous learning, organizations can successfully scale generative AI solutions and unlock their full potential for driving value and efficiency in various domains.

Figure 14

Chapter 13: Future Directions and Emerging Trends

The Evolving Landscape of Generative AI for Enterprise Decision Intelligence:

Unleashing New Possibilities

Enterprise decision intelligence has undergone a transformative journey in recent years, with generative artificial intelligence (AI) emerging as a powerful tool within this landscape. Generative AI, including techniques such as generative adversarial networks (GANs) and variational autoencoders (VAEs), has the ability to generate new content, simulate scenarios, and provide valuable insights to inform decision-making processes.

As the field continues to evolve, generative AI is opening up new possibilities and redefining the way organizations approach decision intelligence. In this section, we will explore the evolving landscape of generative AI for enterprise decision intelligence and its potential implications.

1. Generating Synthetic Data:

One significant application of generative AI in decision intelligence is the generation of synthetic data. Traditional data scarcity and privacy concerns often limit the availability of real-world data. Generative AI models can address this challenge by generating synthetic data that closely resembles real data patterns. This synthetic data can be used to augment existing datasets, create diverse training samples, and overcome data limitations in various domains, such as healthcare, finance, and manufacturing. By leveraging generative AI for synthetic data generation, organizations can enhance the accuracy and robustness of their decision models.

2. Scenario Simulation and What-If Analysis:

Generative AI models enable scenario simulation and what-if analysis, providing decision-makers with valuable insights into potential outcomes and alternative possibilities. Decision intelligence systems powered by generative AI can generate multiple scenarios and simulate the impact of different decisions or external factors. This capability aids in risk assessment, resource allocation, strategic planning, and policy evaluation. Decision-makers can explore various what-if scenarios, understand their implications, and make informed decisions based on comprehensive insights.

3. Creative Content Generation:

Generative AI is increasingly being applied to creative content generation, ranging from art and music to design and marketing. These models can generate novel and diverse content, inspiring innovation and creativity within organizations. For example, generative AI can assist in creating personalized marketing campaigns, designing unique product prototypes, or generating interactive visuals for data exploration. By leveraging generative AI in creative content generation, organizations can streamline creative processes, unlock new design possibilities, and enhance customer engagement.

4. Human-AI Collaboration:

The evolving landscape of generative AI emphasizes the importance of human-AI collaboration. Rather than replacing human expertise, generative AI serves as a powerful tool that augments human decision-making capabilities. Decision intelligence systems that facilitate seamless collaboration between humans and generative AI models enable organizations to leverage the strengths of both. Humans bring domain knowledge, intuition, and contextual understanding, while generative AI contributes with data-driven insights, scenario generation, and pattern recognition. This collaboration leads to more informed and effective decision-making.

5. Ethical Considerations and Fairness:

As generative AI becomes more prevalent in decision intelligence, addressing ethical considerations and ensuring fairness is crucial. Generative AI models can inadvertently amplify biases present in the training data or introduce new biases. Organizations must employ techniques such as bias detection, bias mitigation, and fairness-aware training to mitigate these issues. By actively considering ethical implications and fairness in the development and deployment of generative AI models, organizations can promote inclusive decision-making and avoid perpetuating biases or discriminatory outcomes.

6. Interpretability and Explainability:

Interpretability and explainability are important aspects of generative AI for decision intelligence. Organizations are increasingly seeking ways to understand how generative AI models arrive at their outputs and make decisions. Techniques such as interpretability methods, attention mechanisms, or feature importance analysis can shed light on the inner workings of these models, improving trust, transparency, and accountability. Decision-makers need to comprehend and explain the decision-making process to gain confidence in the outputs and ensure ethical and responsible decision intelligence.

In conclusion, the evolving landscape of generative AI for enterprise decision intelligence holds immense promise. From generating synthetic data to facilitating scenario simulation, creative content generation, human-AI collaboration, and addressing ethical considerations, generative AI is transforming the way organizations approach decision-making processes. By embracing these advancements and leveraging generative AI's capabilities, organizations can gain a competitive edge, make informed decisions, and unlock new possibilities for growth and innovation in an increasingly complex business environment.

The Convergence of Generative AI with Other Technologies (e.g., blockchain, IoT): Expanding Possibilities and Transforming Industries

The field of generative artificial intelligence (AI) has seen significant advancements in recent years, demonstrating its ability to generate new content, simulate scenarios, and provide valuable insights. However, the true power of generative AI is realized when it converges with other transformative technologies. The convergence of generative AI with technologies like blockchain, the Internet of Things (IoT), and others creates a synergistic effect, expanding possibilities and transforming industries. In this section, we will explore the convergence of generative AI with other technologies and examine the potential implications across various domains.

1. Generative AI and Blockchain:
The combination of generative AI and blockchain brings unique opportunities for transparency, trust, and decentralized collaboration. Blockchain's decentralized and immutable nature can ensure the integrity and traceability of generative AI-generated content. Smart contracts on blockchain platforms can facilitate the secure and transparent licensing, ownership, and distribution of generative AI models or creative works. Additionally, blockchain can enhance data privacy by enabling selective sharing of data while preserving ownership and control. The convergence of generative AI with blockchain has the potential to revolutionize content creation, intellectual property management, and secure collaboration across industries.

2. Generative AI and Internet of Things (IoT):
The integration of generative AI with the Internet of Things (IoT) unlocks new possibilities for intelligent automation, predictive analytics, and personalized experiences. By combining real-time data from IoT devices with generative AI models, organizations can create dynamic and adaptive systems. For example, in smart cities, generative AI can analyze IoT data streams to optimize resource allocation, predict traffic patterns, or enhance energy efficiency. In healthcare, generative AI can leverage IoT data to personalize treatment plans and support real-time monitoring. The convergence of generative AI and IoT enables organizations to make data-driven decisions, enhance operational efficiency, and deliver tailored experiences.

3. Generative AI and Robotics:
The convergence of generative AI and robotics amplifies the potential for intelligent automation, human-robot collaboration, and adaptive robotics systems. Generative AI models can enhance robotics capabilities by enabling robots to generate and adapt their behaviors based on real-time data and contextual information. This convergence can lead to the development of robots that can learn, adapt, and interact with humans in more sophisticated ways. For instance, generative AI can enable robots to generate natural language responses, mimic human-like gestures, or adapt

their movements to changing environments. The combination of generative AI and robotics holds significant promise for industries such as manufacturing, healthcare, and logistics.

4. Generative AI and Augmented Reality/Virtual Reality (AR/VR):

The convergence of generative AI with augmented reality (AR) and virtual reality (VR) technologies creates immersive and interactive experiences. Generative AI can generate realistic virtual environments, objects, or characters in AR/VR applications. It enables personalized content generation, dynamic simulations, and interactive experiences that respond to user inputs. For example, generative AI can create virtual characters that exhibit human-like behaviors or generate realistic virtual environments for training simulations. The combination of generative AI with AR/VR technologies enhances the level of immersion, interactivity, and realism in virtual experiences.

5. Generative AI and Natural Language Processing (NLP):

The convergence of generative AI with natural language processing (NLP) technologies enhances language understanding, conversation generation, and content creation. Generative AI models can generate human-like text, engage in meaningful conversations, or assist in content creation tasks. This convergence has broad applications in customer service, chatbots, content generation, and language translation. For instance, generative AI-powered chatbots can provide personalized and contextually relevant responses, improving customer interactions. In content creation, generative AI can assist in generating articles, summaries, or creative writing. The combination of generative AI and NLP enables organizations to automate language-related tasks and enhance communication capabilities.

In conclusion, the convergence of generative AI with other transformative technologies is revolutionizing industries and expanding possibilities. The integration of generative AI with blockchain, IoT, robotics, AR/VR, NLP, and others unlocks new dimensions of transparency, automation, personalization, and immersive experiences. The potential applications span across various domains, including content creation, healthcare, manufacturing, smart cities, and more. Embracing the convergence of generative AI with other technologies enables organizations to achieve new levels of efficiency, innovation, and value creation in an increasingly interconnected world.

Leveraging Generative AI for Strategic Decision-Making and Innovation: Unleashing Creativity and Driving Business Success

In today's rapidly evolving business landscape, organizations face the ongoing challenge of making strategic decisions that can shape their future and drive sustainable growth. To navigate this complex environment, organizations need to leverage advanced technologies that can provide valuable insights, optimize decision-making processes, and foster innovation. Generative artificial intelligence (AI) has emerged as a powerful tool in this regard, offering organizations the ability to make informed decisions, unleash creativity, and drive business success.

Generative AI, which encompasses techniques such as generative adversarial networks (GANs) and variational autoencoders (VAEs), has the unique ability to generate new content, simulate scenarios, and provide valuable insights. By harnessing the potential of generative AI, organizations can overcome traditional decision-making limitations and unlock innovative solutions. In this section, we will explore how generative AI can be leveraged for strategic decision-making and innovation.

1. Scenario Exploration and Risk Assessment:

Generative AI allows decision-makers to simulate and explore various scenarios, enabling them to evaluate potential outcomes and assess risks. By generating multiple scenarios based on historical data or hypothetical inputs, generative AI models provide insights into different possibilities and their associated risks. This capability helps decision-makers make informed choices, anticipate potential challenges, and develop strategies that are robust and adaptable to different scenarios.

2. Data-Driven Insights and Pattern Recognition:

Generative AI models have the ability to extract patterns and insights from vast amounts of data. By training on large datasets, these models can identify hidden patterns, trends, and correlations that may not be apparent to human analysts. Leveraging generative AI for data-driven insights enables organizations to make more accurate predictions, identify emerging market trends, and uncover opportunities for innovation. These insights serve as a foundation for strategic decision-making, enabling organizations to stay ahead of the competition and seize new growth opportunities.

3. Idea Generation and Innovation:

Generative AI models can foster innovation by generating new and creative ideas. Organizations can leverage generative AI to explore uncharted territories, ideate novel solutions, and stimulate creativity. By feeding generative AI models with relevant data and parameters, organizations can generate a diverse range of ideas and concepts. These ideas can spark innovation, inspire new product development, and challenge conventional thinking. Generative AI acts as a catalyst for ideation, fueling a culture of innovation within organizations.

4. Personalization and Customer-Centric Decision-Making:

Generative AI enables organizations to personalize products, services, and experiences to meet the specific needs and preferences of individual customers. By training generative AI models on customer data and behavior patterns, organizations can generate personalized recommendations, tailored marketing campaigns, and customized offerings. This customer-centric approach to decision-making enhances customer satisfaction, loyalty, and drives business growth.

5. Design and Creative Content Generation:

Generative AI models have shown remarkable capabilities in generating design variations and creative content. Organizations can leverage generative AI to automate design processes, generate visual assets, or create interactive experiences. For example, in architecture and fashion, generative AI can generate innovative design concepts and assist in the creative process. In content creation, generative AI can aid in generating articles, artwork, or music compositions. By incorporating generative AI into the creative workflow, organizations can streamline processes, accelerate innovation, and push the boundaries of creativity.

6. Optimization and Decision Support:

Generative AI can optimize decision-making processes by providing decision support and automated decision recommendations. By combining generative AI with optimization algorithms, organizations can identify optimal solutions, allocate resources efficiently, and automate repetitive decision-making tasks. This frees up human decision-makers to focus on higher-level strategic thinking, while generative AI supports them with data-driven insights and recommendations.

7. Rapid Prototyping and Iterative Improvement:

Generative AI enables rapid prototyping and iterative improvement, facilitating agile decision-making and innovation cycles. Organizations can use generative AI to generate prototypes or simulate product variations, allowing for quick evaluation and testing. The feedback received from these prototypes can be used to refine and improve products or solutions iteratively. This iterative approach allows for experimentation, adaptation, and optimization, ultimately leading to the identification of optimal solutions.

In conclusion, leveraging generative AI for strategic decision-making and innovation provides organizations with a powerful toolset to navigate the complexities of the business landscape. By harnessing the potential of generative AI for scenario exploration, data-driven insights, idea generation, personalization, design, optimization, and rapid prototyping, organizations can make informed decisions, drive innovation, and achieve long-term success. Embracing generative AI as a strategic enabler allows organizations to stay ahead of the curve, seize new opportunities, and unleash their creative potential.

Exploring the Boundaries of Creativity and Generative Design in Decision-Making

In the realm of decision-making, creativity plays a crucial role in finding innovative solutions and driving business success. Traditionally, decision-making has relied on human intuition, expertise, and analysis of historical data. However, with the emergence of generative design and artificial intelligence (AI), organizations now have a powerful tool at their disposal to explore the boundaries of creativity and enhance decision-making processes.

Generative design combines the principles of AI, computational algorithms, and human input to generate a multitude of design options based on specified constraints and goals. It enables organizations to explore a vast design space, uncover novel solutions, and push the boundaries of what is possible. By leveraging generative design in decision-making, organizations can unlock new levels of creativity and innovation. In this section, we will delve into the potential of generative design in decision-making and explore how it expands the boundaries of creativity.

1. Amplifying Creativity through Exploration:

Generative design liberates decision-makers from the constraints of conventional thinking by generating a wide range of design options. By exploring a diverse set of possibilities, decision-makers can break away from preconceived notions and discover innovative solutions. Generative design algorithms consider multiple parameters, constraints, and objectives to generate designs that optimize various criteria simultaneously. This exploration of possibilities fuels creativity, leading to breakthrough ideas and unconventional approaches to problem-solving.

2. Optimal Solutions through Iterative Improvement:

Generative design supports an iterative design process, enabling decision-makers to refine and improve designs over time. By generating multiple design iterations, decision-makers can evaluate, select, and refine the most promising options. Feedback from each iteration informs the subsequent generation of designs, resulting in continuous improvement. This iterative approach allows for experimentation, adaptation, and optimization, ultimately leading to the identification of optimal solutions.

3. Fostering Collaborative Decision-Making:

Generative design facilitates collaborative decision-making by providing a platform for stakeholders from various disciplines to contribute their expertise. Architects, engineers, designers, and other stakeholders can collaborate and provide input during the generative design process. By incorporating diverse perspectives, knowledge, and insights, organizations can make more informed decisions that consider a range of factors and trade-offs. This collaborative approach promotes interdisciplinary thinking and fosters a culture of innovation within organizations.

4. Enabling Data-Driven Design Decisions:

Generative design relies on data and analytics to inform the design generation process. By leveraging historical data, market trends, and user feedback, generative design algorithms can generate designs that align with customer preferences and market demands. This data-driven approach ensures that design decisions are rooted in empirical evidence and user-centric considerations. Organizations can make informed decisions backed by data-driven insights, reducing the reliance on intuition and guesswork.

5. Overcoming Design Constraints and Pushing Boundaries:

Generative design empowers decision-makers to overcome design constraints and push the boundaries of what is traditionally deemed possible. By defining specific constraints, such as material properties, structural requirements, or manufacturing limitations, generative design algorithms can optimize designs within these boundaries. This allows decision-makers to explore unconventional materials, forms, and configurations that challenge traditional design norms. Generative design opens up new avenues for creativity, innovation, and problem-solving.

6. Enhancing Efficiency and Speed:

Generative design expedites the decision-making process by automating repetitive tasks and accelerating design iterations. Traditional design processes often involve manual iterations, time-consuming modifications, and trial-and-error approaches. Generative design algorithms streamline these processes, automating design generation, evaluation, and optimization. This efficiency allows decision-makers to explore a broader design space, consider more options, and arrive at decisions faster, saving time and resources.

In conclusion, the integration of generative design into decision-making processes expands the boundaries of creativity, enabling organizations to explore a broader design space, discover innovative solutions, and optimize designs based on multiple criteria. By amplifying creativity, fostering collaboration, enabling data-driven decisions, and overcoming design constraints, generative design empowers decision-makers to make more informed and innovative choices. Embracing generative design in decision-making processes can drive breakthrough innovations, enhance user experiences, and position organizations at the forefront of their industries.

Ethical Considerations and Responsible AI in Enterprise Decision Intelligence: Building Trust and Accountability

In the age of data-driven decision-making, organizations increasingly rely on artificial intelligence (AI) and machine learning (ML) algorithms to derive insights, optimize processes, and inform strategic choices. However, as AI continues to shape the enterprise decision intelligence landscape, ethical considerations and responsible AI practices become paramount. Organizations must navigate the ethical complexities to ensure transparency, fairness, and accountability in their AI-driven decision-making processes. In this section, we will explore the ethical considerations and responsible AI practices that are crucial in the realm of enterprise decision intelligence.

1. Transparency and Explainability:

One of the key ethical considerations in decision intelligence is the transparency and explainability of AI models. Organizations should strive to understand and communicate how AI models arrive at their decisions, especially when those decisions have significant impacts on individuals or society. Techniques such as interpretable AI, explainable AI, and model-agnostic approaches can shed light on the decision-making process, providing insights into the factors and features that influence outcomes. Transparent and explainable AI instills trust, enables better auditability, and ensures that decisions are made in a fair and accountable manner.

2. Fairness and Bias Mitigation:

Ensuring fairness in decision-making processes is critical for responsible AI implementation. AI models can inadvertently perpetuate biases present in training data, leading to discriminatory outcomes. Organizations must actively identify and mitigate biases in their AI systems. This involves careful consideration of data collection, training data selection, and evaluation metrics. Techniques such as fairness-aware training, bias detection, and algorithmic audits can help organizations identify and address bias in AI models. By striving for fairness and equity, organizations can avoid discriminatory practices and promote inclusivity in decision intelligence.

3. Privacy and Data Protection:

Responsible AI implementation requires organizations to prioritize privacy and data protection. Decision intelligence often relies on large volumes of sensitive data, which must be handled with utmost care. Organizations must comply with data protection regulations, ensure data anonymization where necessary, and implement appropriate security measures to safeguard personal information. Privacy-enhancing technologies, differential privacy, and federated learning techniques can be employed to strike a balance between data utility and privacy preservation. By respecting privacy rights and maintaining data security, organizations can foster trust and confidence among stakeholders.

4. Ethical Decision-Making Frameworks:

Developing and implementing ethical decision-making frameworks is essential for responsible AI in enterprise decision intelligence. Organizations should establish guidelines and principles that align with ethical standards, legal requirements, and societal expectations. These frameworks should be well-defined, widely communicated, and integrated into the decision-making process. Stakeholder engagement and multidisciplinary collaboration can help identify ethical considerations and ensure that AI-driven decisions are aligned with organizational values and societal norms.

5. Human Oversight and Intervention:

While AI plays a crucial role in decision intelligence, human oversight and intervention remain essential. Human judgment, experience, and ethical reasoning are invaluable in assessing AI outputs, evaluating potential risks, and addressing complex ethical dilemmas. Organizations should maintain human accountability, ensuring that AI is used as a tool to augment human decision-making rather than replacing it. Human-in-the-loop approaches, human-AI collaboration, and clear delineation of responsibilities between humans and AI systems foster responsible decision intelligence practices.

6. Continuous Monitoring and Evaluation:

Responsible AI practices require ongoing monitoring and evaluation of AI systems. Organizations should regularly assess the performance, fairness, and societal impact of their AI models. This includes conducting regular audits, soliciting feedback from stakeholders, and tracking the outcomes of AI-driven decisions. Continuous monitoring allows organizations to detect and address any biases, errors, or ethical concerns that may arise. It enables organizations to iteratively improve their AI systems, ensuring they align with ethical considerations and contribute positively to decision intelligence.

In conclusion, ethical considerations and responsible AI practices are imperative in enterprise decision intelligence. Organizations must prioritize transparency, fairness, privacy protection, and human oversight to build trust, ensure accountability, and mitigate the risks associated with AI-driven decision-making. By integrating ethical considerations into their decision intelligence frameworks, organizations can foster a culture of responsible AI, make ethical choices, and leverage AI's potential for positive societal impact. Ultimately, ethical decision intelligence lays the foundation for sustainable success and helps organizations navigate the complex ethical landscape of the modern world.

"Open wide and say A-I-I-I-I-I-I-I-I-I-I."

Figure 15

Chapter 14: Apply Generative AI for Enterprise Decision Intelligence to Industries

Applying to Insurance Industry including a Business Case, Case Study, and SWOT Analysis

Revolutionizing the Insurance Industry through Enterprise Strategy Intelligence

Introduction: The insurance industry operates in a complex landscape, with evolving customer expectations, dynamic regulatory environments, and emerging technologies reshaping the industry. In order to thrive in this rapidly changing environment, insurance companies need to adopt strategic approaches that enable them to make informed decisions and drive sustainable growth. One such approach is the application of Enterprise Strategy Intelligence (ESI). By leveraging the power of ESI, insurance companies can gain a competitive edge, optimize operations, and deliver enhanced value to customers. In this section, we will explore the transformative potential of ESI when applied to the insurance industry.

Understanding Enterprise Strategy Intelligence: Enterprise Strategy Intelligence (ESI) involves the systematic collection, analysis, and interpretation of data to inform strategic decision-making. It encompasses leveraging advanced technologies, data analytics, and industry insights to gain a comprehensive understanding of internal and external factors influencing the insurance business. By harnessing ESI, insurance companies can effectively respond to market dynamics, identify emerging trends, mitigate risks, and seize growth opportunities.

Enhancing Customer Engagement and Experience: In an era where customer expectations are continually rising, insurers must prioritize delivering exceptional customer experiences. ESI enables insurers to gather and analyze vast amounts of customer data, allowing them to gain actionable insights into customer preferences, behavior patterns, and needs. By leveraging these insights, insurers can tailor their products, services, and communication strategies to meet individual customer requirements. ESI also enables the implementation of personalized digital channels, streamlined claims processes, and proactive customer service, resulting in improved customer satisfaction and long-term loyalty.

Optimizing Underwriting and Risk Management: Accurate risk assessment and management are at the heart of the insurance industry. ESI empowers insurers with advanced analytics and predictive models to optimize underwriting processes and enhance risk management capabilities. By leveraging historical data, market trends, and real-time information, insurers can make data-driven decisions regarding policy pricing, coverage limits, and risk diversification. ESI also facilitates the identification of emerging risks, enabling insurers to develop innovative insurance products that address evolving market needs.

Enabling Data-Driven Decision Making: Insurance companies deal with vast amounts of data from multiple sources, including policyholders, claims, and market trends. ESI equips insurers with the tools and technologies to effectively collect, analyze, and interpret this data, transforming it into actionable insights. By leveraging advanced analytics and artificial intelligence, insurers can uncover hidden patterns, detect fraud, and improve claims processing efficiency. ESI also enables insurers to gain a comprehensive view of their operations, identify areas of improvement, and optimize resource allocation.

Navigating Regulatory Compliance: The insurance industry is subject to complex and evolving regulatory frameworks. ESI helps insurers navigate these regulatory challenges by monitoring and analyzing changes in regulations, compliance requirements, and industry standards. By leveraging ESI, insurers can proactively adapt their strategies, ensure compliance, and avoid potential penalties or reputational risks. ESI also enables insurers to stay abreast of emerging compliance trends and incorporate them into their business practices.

Conclusion: As the insurance industry continues to evolve, embracing Enterprise Strategy Intelligence becomes crucial for success. By leveraging advanced analytics, market intelligence, and data-driven decision-making, insurers can gain a competitive advantage, optimize operations, and deliver superior customer experiences. ESI empowers insurance companies to make informed strategic choices, enhance risk management capabilities, and navigate regulatory challenges. By harnessing the power of ESI, insurers can position themselves as industry leaders, drive innovation, and achieve sustainable growth in the rapidly evolving insurance landscape.

Business Case for Enterprise Strategy Intelligence for the Insurance Industry	
Case	Action
Executive Summary:	The insurance industry operates in a highly competitive and rapidly evolving market, facing challenges such as changing customer expectations, regulatory changes, emerging technologies, and new entrants. To maintain a competitive edge, insurance companies need to make informed strategic decisions based on accurate and timely insights. Implementing an Enterprise Strategy Intelligence (ESI) system will enable insurance companies to gather, analyze, and leverage data-driven insights, empowering them to develop and execute effective strategies, optimize operations, and achieve sustainable growth
Business Objectives:	**Enhance Strategic Decision-Making:** ESI enables insurance companies to gain a comprehensive view of their market, competitors, and customer segments. This intelligence equips decision-makers with valuable insights to identify new growth opportunities, develop effective product offerings, and allocate resources strategically. **Improve Operational Efficiency:** By leveraging ESI, insurance companies can optimize operational processes, streamline workflows, and identify areas for cost reduction. Through data-driven insights, companies can improve underwriting accuracy, claims management, fraud detection, and risk assessment, resulting in enhanced efficiency and reduced operational costs. **Enhance Customer Experience:** ESI provides insurance companies with a deeper understanding of customer behavior, preferences, and needs. By leveraging this intelligence, companies can develop personalized products, tailor customer interactions, and provide proactive services, thereby improving customer satisfaction, retention, and loyalty. **Mitigate Risks:** ESI equips insurance companies with real-time risk intelligence, enabling proactive risk management and mitigation strategies. By leveraging predictive analytics, companies can identify potential risks, assess their impact, and develop appropriate risk mitigation plans, leading to improved underwriting accuracy, reduced claims costs, and enhanced portfolio management.

Benefits and Value Proposition:	**Strategic Insights:** ESI provides insurance companies with comprehensive market intelligence, enabling them to identify emerging trends, market gaps, and potential opportunities. This empowers decision-makers to make informed and proactive strategic choices, leading to competitive advantages and revenue growth. **Operational Efficiency:** By leveraging ESI, insurance companies can optimize operational processes, reduce manual efforts, and enhance productivity. Automation of routine tasks, data analysis, and reporting enables employees to focus on higher-value activities, improving overall operational efficiency. **Risk Mitigation:** ESI enables insurance companies to proactively identify and manage risks. By leveraging data analytics and predictive models, companies can identify potential risks, develop risk mitigation strategies, and optimize risk portfolios, leading to reduced losses, improved underwriting accuracy, and enhanced profitability. **Customer-Centric Approach:** ESI empowers insurance companies to understand their customers better, enabling personalized product offerings, tailored communication, and superior customer service. This customer-centric approach enhances customer satisfaction, loyalty, and retention, ultimately driving revenue growth.

Implementation Plan:	**Assess Needs and Objectives:** Conduct a thorough assessment of the organization's strategic goals, operational challenges, and data requirements. Identify key stakeholders and involve them in the decision-making process.
	Technology Infrastructure: Evaluate and select a robust ESI platform that aligns with the organization's needs, scalability requirements, and data integration capabilities. Ensure the platform supports advanced analytics, data visualization, and real-time reporting.
	Data Integration and Governance: Establish a data governance framework to ensure data quality, security, and compliance. Integrate internal and external data sources, such as customer data, market data, claims data, and regulatory information, into a centralized data repository.
	Analytics Capabilities: Develop analytical models, algorithms, and dashboards to extract insights from the integrated data. Leverage techniques such as machine learning, natural language processing, and predictive analytics to generate actionable intelligence.
	Change Management: Implement a comprehensive change management plan to ensure successful adoption of the ESI system. Train employees on data-driven decision-making, promote a data-driven culture, and align performance metrics with strategic objectives.
	Continuous Improvement: Establish feedback loops and monitor key performance indicators to measure the effectiveness of the ESI system. Continuously refine and improve the system based on user feedback, market changes, and emerging technologies

Cost and ROI Analysis:	**Upfront Investment:** Initial costs include software licensing, hardware infrastructure, data integration, and implementation services. Additionally, there will be costs associated with change management, training, and data governance.
	Cost Reduction: ESI enables operational efficiencies, reducing costs associated with manual processes, claims handling, and risk management. By optimizing underwriting accuracy, claims management, and fraud detection, companies can minimize losses and improve profitability.
	Revenue Growth: ESI empowers insurance companies to identify new growth opportunities, develop customer-centric products, and enhance customer retention. This leads to increased policy sales, improved cross-selling, and revenue growth.
	Competitive Advantage: ESI provides insurance companies with a competitive edge by enabling data-driven decision-making and strategic insights. This advantage can lead to market share gains, improved customer satisfaction, and a stronger market position.

Overall, the implementation of an Enterprise Strategy Intelligence system in the insurance industry offers substantial benefits, including enhanced strategic decision-making, operational efficiency, risk mitigation, and customer experience. By leveraging data-driven insights, insurance companies can thrive in a dynamic market, achieve sustainable growth, and stay ahead of the competition.

ROI Analysis: Applying Enterprise Strategy Intelligence to the Insurance Industry

Enterprise Strategy Intelligence (ESI) holds the potential to generate significant returns on investment for insurance companies. By leveraging advanced data analytics, machine learning, and strategic decision-making, ESI can drive operational efficiency, enhance risk assessment, improve customer experiences, and increase profitability. Let's explore the potential return on investment (ROI) of applying ESI in the insurance industry.

ROI Analysis: Applying Enterprise Strategy Intelligence to the Insurance Industry	
Case	Action
Operational Efficiency:	ESI streamlines internal processes, reduces manual effort, and optimizes resource allocation. This leads to improved operational efficiency, cost savings, and increased productivity. By automating tasks such as claims processing, underwriting, and policy administration, insurance companies can reduce administrative overheads, minimize errors, and accelerate time-to-market. The ROI from improved operational efficiency includes reduced operational costs, increased throughput, and enhanced customer satisfaction.
Risk Assessment and Management:	ESI enables insurance companies to enhance their risk assessment capabilities, resulting in more accurate pricing, underwriting, and claims management. By leveraging advanced analytics and predictive modeling techniques, insurers can better understand risks, identify potential fraud, and improve loss ratio management. The ROI from improved risk assessment includes reduced claim payouts, minimized losses from fraudulent activities, and optimized underwriting practices.
Customer Experience and Retention:	ESI allows insurers to personalize customer experiences, leading to improved customer satisfaction, loyalty, and retention. By analyzing customer data, preferences, and behavior, insurers can tailor their offerings, provide proactive recommendations, and deliver targeted marketing campaigns. This leads to increased customer engagement, higher cross-selling and upselling opportunities, and reduced customer churn. The ROI from enhanced customer experiences includes increased policy renewals, higher customer lifetime value, and improved brand loyalty.

ROI Analysis: Applying Enterprise Strategy Intelligence to the Insurance Industry	
Product Innovation and Market Differentiation:	ESI provides insurers with valuable insights into market trends, emerging risks, and customer demands. This information can drive product innovation and help insurers develop unique offerings that cater to specific market segments. By leveraging ESI to identify underserved market niches, insurers can develop differentiated products, gain a competitive edge, and capture new market share. The ROI from product innovation and market differentiation includes increased market penetration, higher premium revenue, and improved market positioning.
Fraud Detection and Prevention:	ESI's advanced analytics capabilities enable insurers to detect and prevent fraudulent activities more effectively. By leveraging machine learning algorithms and anomaly detection techniques, insurers can identify patterns indicative of fraud, automate fraud detection processes, and reduce financial losses. The ROI from improved fraud detection and prevention includes reduced claims fraud losses, improved loss ratio, and enhanced brand reputation.
Strategic Decision-Making:	ESI empowers insurers with data-driven insights and advanced analytics to support strategic decision-making. By leveraging comprehensive data analysis, predictive modeling, and scenario simulation, insurers can make informed decisions that optimize business outcomes. This leads to improved profitability, better investment decisions, and enhanced strategic planning. The ROI from strategic decision-making includes increased profitability, optimized investment returns, and improved market position.

Calculating the exact ROI of implementing ESI will depend on factors such as the scale of implementation, organizational readiness, and specific business objectives. However, the potential benefits outlined above demonstrate the positive impact ESI can have on insurance companies. To maximize ROI, organizations should conduct a thorough analysis of their unique needs, set clear goals, and implement ESI in a strategic and well-executed manner.

In conclusion, applying Enterprise Strategy Intelligence (ESI) to the insurance industry offers a compelling ROI. By driving operational efficiency, improving risk assessment, enhancing customer experiences, fostering product innovation, preventing fraud, and supporting strategic decision-making, ESI enables insurance companies to achieve sustainable growth, competitive advantage, and increased profitability.

SWOT Analysis: Applying Enterprise Strategy Intelligence to the Insurance Industry

SWOT ANALYSIS — **INTERNAL**

POSITIVE / **NEGATIVE**

STRENGTHS

1. Enhanced Decision-Making: Enterprise Strategy Intelligence (ESI) empowers insurance companies with data-driven insights, enabling more informed and strategic decision-making. By analyzing vast amounts of structured and unstructured data, ESI can identify market trends, customer preferences, and emerging risks, helping insurance companies stay ahead of the competition.

2. Improved Risk Assessment: ESI enables insurance companies to assess risks more accurately by leveraging advanced analytics and predictive modeling techniques. By analyzing historical data, market conditions, and external factors, ESI can provide a comprehensive understanding of risks, enabling insurers to better price policies, manage underwriting processes, and minimize potential losses.

3. Personalized Customer Experiences: ESI allows insurance companies to tailor their products and services to individual customer needs. By leveraging data analytics, machine learning, and customer segmentation techniques, ESI can help insurers identify customer preferences, anticipate future needs, and offer personalized coverage options. This enhances customer satisfaction, retention, and loyalty.

4. Efficient Operations: ESI streamlines internal processes and operational efficiency within insurance companies. By automating manual tasks, optimizing resource allocation, and improving workflow management, ESI enhances operational productivity and reduces costs. This allows insurers to allocate resources strategically, optimize claims processing, and improve overall efficiency.

WEAKNESSES

1. Data Privacy and Security Concerns: The application of ESI in the insurance industry raises concerns about data privacy and security. Insurers handle sensitive customer information, and the collection and analysis of large volumes of data introduce potential vulnerabilities. Ensuring robust data privacy measures, complying with regulations, and implementing strong security protocols are essential to address this weakness.

2. Dependence on Data Quality: ESI relies heavily on accurate and high-quality data for reliable decision-making. Inaccurate or incomplete data can lead to flawed insights and misguided decisions. Insurers must establish robust data governance frameworks, invest in data cleansing and validation processes, and address data quality issues to mitigate this weakness.

---

1. Enhanced Fraud Detection and Prevention: ESI can be utilized to identify patterns and anomalies in data, aiding in fraud detection and prevention. By leveraging advanced analytics and machine learning algorithms, insurers can proactively detect fraudulent activities, improving risk management and reducing financial losses.

2. Product Innovation and Market Differentiation: ESI provides insurers with insights into market trends, customer preferences, and emerging risks. This information can guide product innovation and help insurers develop unique offerings that cater to specific customer needs. By leveraging ESI, insurers can differentiate themselves in the market and gain a competitive edge.

---

1. Technological Challenges: Implementing ESI requires robust technological infrastructure, skilled personnel, and integration with existing systems. The complexity of integrating data sources, managing large datasets, and ensuring data accuracy poses challenges for insurers. Addressing these technological challenges through proper planning, investment, and training is crucial to successfully implement ESI.

2. Regulatory and Compliance Requirements: The insurance industry is subject to strict regulations and compliance requirements. The application of ESI must adhere to these regulations, ensuring transparency, fairness, and ethical practices. Failure to comply with regulatory frameworks can lead to legal consequences and reputational damage.

3. Resistance to Change: Introducing ESI into traditional insurance organizations may face resistance from employees and stakeholders who are unfamiliar with or resistant to change. Overcoming resistance requires effective change management strategies, clear communication, and fostering a culture of data-driven decision-making.

OPPORTUNITIES

THREATS

EXTERNAL

Case Study: Enterprise Strategy Intelligence for Large Midwest Insurance Company

Case	Action
Copmany Overview	Large Midwest Insurance Company, a leading player in the insurance industry, recognized the need to gain a competitive edge in a rapidly evolving market. To achieve their strategic objectives, they implemented an Enterprise Strategy Intelligence (ESI) system. This case study highlights the benefits and outcomes of Large Midwest Insurance Company's ESI implementation.
Business Objectives:	**Enhance Strategic Decision-Making:** Large Midwest Insurance Company aimed to gain comprehensive market insights, identify growth opportunities, and develop effective strategies to stay ahead of the competition. **Improve Operational Efficiency:** The company sought to streamline operational processes, optimize underwriting accuracy, claims management, and risk assessment, resulting in improved efficiency and reduced costs. **Enhance Customer Experience:** Large Midwest Insurance Company aimed to deepen their understanding of customer preferences and behaviors to offer personalized products, tailored services, and exceptional customer experiences. **Mitigate Risks:** The company aimed to proactively identify and mitigate risks by leveraging data analytics and predictive models, leading to improved underwriting accuracy and reduced claims costs.

| Implementation Process: | **Needs Assessment:** Large Midwest Insurance Company conducted a thorough assessment of their strategic goals, operational challenges, and data requirements. They identified key stakeholders, including executives, underwriters, claims specialists, and IT professionals, to ensure alignment throughout the implementation process.

Technology Infrastructure: After careful evaluation, Large Midwest Insurance Company selected an ESI platform that offered advanced analytics capabilities, robust data integration, and real-time reporting. The platform was customized to meet their specific needs and integrated seamlessly with existing systems.

Data Integration and Governance: Large Midwest Insurance Company established a data governanceo framework to ensure data quality, security, and compliance. They integrated internal and external data sources, including customer data, market data, claims data, and regulatory information, into a centralized data repository.

Analytics Capabilities: Large Midwest Insurance Company developed analytical models, algorithms, and interactive dashboards to extract actionable insights from the integrated data. They leveraged machine learning techniques to predict customer behavior, identify potential risks, and optimize underwriting decisions.

Change Management: A comprehensive change management plan was implemented to foster a data-driven culture within the organization. Employees received training on the ESI system, data-driven decision-making, and the benefits of leveraging intelligence for strategic initiatives. |
|---|---|

Outcomes and Benefits:	**Enhanced Strategic Decision-Making:** With the ESI system in place, Large Midwest Insurance Company gained real-time market intelligence, enabling them to identify emerging trends, market gaps, and new growth opportunities. This empowered their decision-makers to make informed and proactive strategic choices, resulting in increased market share and revenue growth. **Improved Operational Efficiency:** By leveraging ESI, Large Midwest Insurance Company optimized operational processes, reducing manual efforts, and enhancing productivity. Automation of routine tasks, data analysis, and reporting freed up valuable resources and enabled employees to focus on high-value activities. Underwriting accuracy and claims management improved, leading to reduced costs and enhanced efficiency. **Enhanced Customer Experience:** ESI provided Large Midwest Insurance Company with a deeper understanding of customer preferences, allowing them to offer personalized products and tailored communication. By leveraging customer intelligence, the company improved customer satisfaction, retention, and loyalty, resulting in increased policy sales and revenue growth. **Effective Risk Mitigation:** The ESI system enabled Large Midwest Insurance Company to proactively identify and manage risks. By leveraging predictive analytics, they accurately assessed potential risks,k developed risk mitigation strategies, and optimized their risk portfolios. This led to improved underwriting accuracy, reduced claims costs, and enhanced profitability.
Conclusion:	The implementation of Enterprise Strategy Intelligence (ESI) transformed Large Midwest Insurance Company's operations, enabling them to make data-driven strategic decisions, optimize efficiency, enhance the customer experience, and mitigate risks effectively. By leveraging the power of intelligence and advanced analytics, Large Midwest Insurance Company gained a competitive advantage, achieving sustainable growth, and maintaining their position as a leading player in the insurance industry.

Applying to Healthcare industry, including a business case, ROI, SWOT Analysis and Implementation Plan, and Case Study

Unleashing the Potential of Enterprise Strategy Intelligence in the Healthcare Industry

The healthcare industry is at the precipice of a transformative era, driven by rapid technological advancements and the increasing complexity of healthcare delivery. At the heart of this transformation lies **Enterprise Strategy Intelligence (ESI)**, a powerful approach that leverages data, analytics, and strategic insights to optimize decision-making and improve outcomes. As the industry grapples with challenges such as rising costs, regulatory pressures, and the need for personalized care, ESI offers a pathway to enhanced efficiency, patient satisfaction, and innovation.

Understanding Enterprise Strategy Intelligence

Enterprise Strategy Intelligence refers to the integration of data analytics, strategic planning, and business intelligence to inform and guide organizational decision-making. In the context of healthcare, ESI encompasses the use of advanced technologies such as artificial intelligence (AI), machine learning (ML), and big data analytics to process vast amounts of information and generate actionable insights. These insights help healthcare organizations streamline operations, improve patient care, and navigate the competitive and regulatory landscape.

Key Components of ESI in Healthcare

1. Data Integration and Management:
The foundation of ESI lies in the ability to integrate and manage diverse data sources, including electronic health records (EHRs), patient surveys, financial reports, and external datasets. Effective data integration ensures that healthcare providers have a comprehensive view of patient information and organizational performance.

2. Predictive Analytics:
Predictive analytics leverages historical data and machine learning algorithms to forecast future trends and outcomes. In healthcare, predictive analytics can anticipate patient admissions, identify high-risk patients, and optimize resource allocation. For instance, predicting patient readmissions enables proactive interventions that can reduce hospital stays and improve patient outcomes.

3. Real-time Monitoring and Reporting:
Real-time monitoring systems provide immediate insights into various aspects of healthcare delivery, from patient vital signs to operational efficiency. These systems enable healthcare providers to respond quickly to emerging issues, such as unexpected patient deterioration or equipment failures, thereby enhancing patient safety and operational resilience.

4. Strategic Planning and Scenario Analysis:
ESI supports strategic planning by simulating different scenarios and assessing their potential impact. This capability allows healthcare leaders to evaluate the effects of policy changes, market

shifts, and technological innovations, ensuring that strategic decisions are well-informed and aligned with organizational goals.

5. Patient-Centric Care Models:

By harnessing data on patient preferences, behaviors, and outcomes, ESI facilitates the development of personalized care models. These models aim to improve patient engagement, adherence to treatment plans, and overall satisfaction. For example, personalized treatment plans based on genetic information and lifestyle factors can lead to more effective and efficient care.

Benefits of ESI in the Healthcare Industry

1. Improved Patient Outcomes:

ESI enables healthcare providers to deliver more accurate diagnoses, tailored treatments, and timely interventions, leading to better patient outcomes. Enhanced data insights allow for a deeper understanding of patient needs and more effective management of chronic conditions.

2. Operational Efficiency:

By optimizing resource allocation and streamlining processes, ESI reduces waste and lowers operational costs. For instance, predictive analytics can help manage staffing levels and ensure that medical supplies are adequately stocked, avoiding shortages and overages.

3. Regulatory Compliance:

The healthcare industry is heavily regulated, with stringent requirements for data privacy, patient safety, and quality of care. ESI helps organizations maintain compliance by providing comprehensive reporting and monitoring tools that ensure adherence to regulatory standards.

4. Innovation and Competitive Advantage:

Organizations that leverage ESI are better positioned to innovate and stay ahead of competitors. By continuously analyzing market trends and technological advancements, healthcare providers can identify new opportunities for growth and improvement.

Challenges and Considerations

While the potential of ESI is vast, its implementation comes with challenges. Data privacy and security are paramount, as healthcare data is highly sensitive. Ensuring that data integration and analytics platforms comply with regulations such as HIPAA is critical. Additionally, the successful adoption of ESI requires a cultural shift within organizations, fostering a data-driven mindset and investing in the necessary technology and training.

Conclusion

Enterprise Strategy Intelligence represents a powerful tool for transforming the healthcare industry. By integrating advanced analytics, strategic planning, and real-time monitoring, ESI empowers healthcare organizations to make informed decisions, improve patient care, and enhance operational efficiency. As the industry continues to evolve, embracing ESI will be key to navigating the complexities of modern healthcare and unlocking new opportunities for innovation and growth.

Business Case: Enterprise Decision Intelligence with Generative AI for the Healthcare Industry

Executive Summary: Describe the key points of the business case, including the need for enterprise strategy intelligence, its potential benefits, and the expected return on investment.

Case	Action
Introduction:	The healthcare industry operates in a complex environment with numerous stakeholders, vast amounts of data, and critical decision-making processes. To improve patient outcomes, enhance operational efficiency, and drive innovation, healthcare organizations must harness the power of data-driven insights and advanced technologies. **Enterprise Decision Intelligence (EDI)** integrated with Generative AI offers a transformative solution that can revolutionize decision-making in the healthcare industry. This business case explores the benefits and potential return on investment (ROI) of implementing EDI with Generative AI in the healthcare sector.
Problem Statement:	The healthcare industry faces challenges related to patient care quality, operational efficiency, cost containment, and medical research advancements. Decision-makers often encounter data overload, information fragmentation, and limited resources, hindering their ability to make well-informed decisions that positively impact patient outcomes.
Solution:	Enterprise Decision Intelligence (EDI) integrated with Generative AI presents a powerful solution to address the healthcare industry's challenges. By leveraging the vast amount of healthcare data available, EDI with Generative AI enables improved clinical decision-making, optimized resource allocation, personalized patient care, and accelerated medical research.

Benefits and ROI:	**a. Clinical Decision Support:** EDI with Generative AI can analyze patient data, medical literature, and treatment guidelines to provide real-time decision support to clinicians. This enhances diagnostic accuracy, treatment selection, and patient safety, resulting in improved patient outcomes and reduced healthcare costs.
	b. Resource Optimization: By leveraging predictive modeling and optimization algorithms, EDI with Generative AI can optimize resource allocation, including staff scheduling, bed utilization, and inventory management. This improves operational efficiency, reduces wait times, and increases patient satisfaction.
	c. Personalized Patient Care: EDI with Generative AI enables the analysis of diverse patient data, including genomics, electronic health records, and wearable device data. This allows for personalized treatment plans, precision medicine approaches, and proactive health management. Personalized patient care improves outcomes, patient satisfaction, and loyalty.
	d. Medical Research and Drug Development: EDI with Generative AI accelerates medical research and drug discovery processes. By analyzing large-scale datasets, identifying patterns, and simulating scenarios, EDI facilitates hypothesis generation, virtual clinical trials, and target identification. This expedites the development of new treatments, reduces costs, and improves the success rate of clinical trials.
	e. Fraud Detection and Regulatory Compliance: EDI with Generative AI assists healthcare organizations in detecting and preventing fraud, waste, and abuse. By analyzing claims data, patient records, and anomalies, EDI identifies irregularities and potential fraudulent activities. This improves regulatory compliance, reduces financial losses, and protects the integrity of healthcare systems.

Implementation Plan:	**a. Data Integration and Infrastructure:** Establish a robust data integration framework to unify disparate data sources, including electronic health records, medical imaging, genomics, and research databases. Invest in secure and scalable data storage and management infrastructure. **b. AI and Generative Models:** Develop and train AI models, leveraging Generative AI techniques such as generative adversarial networks (GANs) and variational autoencoders (VAEs), to generate insights, simulate scenarios, and support decision-making processes. **c. Interdisciplinary Collaboration:** Foster collaboration among healthcare professionals, data scientists, researchers, and IT teams to ensure effective implementation of EDI with Generative AI. Encourage knowledge sharing and interdisciplinary projects to maximize the potential impact of AI-driven decision intelligence. **d. Data Privacy and Security:** Implement robust data privacy and security measures to protect patient information and comply with regulations such as HIPAA. Ensure data anonymization and enforce access controls to maintain patient confidentiality. **e. Change Management and Training:** Develop a comprehensive change management plan to address organizational culture, workflows, and skill gaps. Provide training and education on EDI with Generative AI to healthcare professionals, emphasizing the value of data-driven decision-making and fostering a culture of continuous learning.
Financial Considerations:	**a. Investment:** The implementation of EDI with Generative AI requires initial investment in technology infrastructure, data integration, AI model development, and employee training. **b. ROI:** The ROI of EDI with Generative AI can be measured through various metrics, including improved patient outcomes, reduced healthcare costs, increased operational efficiency, enhanced research productivity, and improved regulatory compliance.

Conclusion:	Implementing Enterprise Decision Intelligence with Generative AI in the healthcare industry can revolutionize decision-making processes, improve patient care, enhance operational efficiency, and drive innovation. By leveraging data-driven insights, predictive modeling, and personalized patient care, healthcare organizations can deliver better outcomes, reduce costs, and stay at the forefront of medical advancements. Embracing EDI with Generative AI positions healthcare organizations for success in an increasingly data-driven and patient-centric industry

ROI Analysis: Applying Enterprise Decision Intelligence with Generative AI to the Healthcare Industry

Introduction:

Applying Enterprise Decision Intelligence (EDI) with Generative AI to the healthcare industry has the potential to transform decision-making processes, improve patient outcomes, enhance operational efficiency, and drive innovation. This ROI analysis explores the financial benefits and potential return on investment (ROI) of implementing EDI with Generative AI in the healthcare sector.

Case	Action
Cost Savings:	**a. Operational Efficiency:** EDI with Generative AI streamlines healthcare operations, reduces manual effort, and optimizes resource allocation. This leads to cost savings through improved productivity, reduced administrative overheads, and optimized workflows. **b. Fraud Detection:** By leveraging AI models and anomaly detection techniques, EDI with Generative AI helps detect and prevent healthcare fraud, waste, and abuse. This results in substantial cost savings by reducing financial losses and preserving the integrity of healthcare systems.
Improved Patient Outcomes:	**a. Clinical Decision Support:** EDI with Generative AI provides real-time decision support to healthcare professionals, improving diagnostic accuracy and treatment selection. This leads to better patient outcomes, reduced readmission rates, and lower healthcare costs associated with ineffective or inappropriate treatments. **b. Personalized Patient Care:** EDI enables personalized treatment plans, precision medicine approaches, and proactive health management. This improves patient outcomes, reduces hospitalization rates, and enhances patient satisfaction and loyalty.
Operational Efficiency:	**a. Resource Optimization:** EDI with Generative AI optimizes resource allocation, including staff scheduling, bed utilization, and inventory management. This leads to improved operational efficiency, reduced wait times, and increased patient throughput. **b. Streamlined Workflows:** EDI automates manual tasks, reduces paperwork, and improves data accessibility. This improves workflow efficiency, enhances collaboration among healthcare professionals, and reduces administrative burdens.

Case	Action
Research and Development:	**a. Accelerated Research:** EDI with Generative AI expedites medical research and drug development processes by analyzing vast amounts of data, identifying patterns, and simulating scenarios. This accelerates the discovery of new treatments, reduces costs, and improves the success rate of clinical trials. **b. Cost Reduction in R&D:** EDI with Generative AI reduces the cost of drug development by enabling virtual trials, identifying potential candidates early in the process, and optimizing trial design. This results in significant cost savings and faster time-to-market for new therapies.
Regulatory Compliance:	EDI with Generative AI assists healthcare organizations in complying with regulatory requirements, such as HIPAA. By automating compliance checks, monitoring data privacy and security, and identifying potential non-compliance issues, EDI mitigates the risk of regulatory penalties and reputational damage. **Calculation of ROI:** To calculate the ROI, consider the following: **a. Initial Investment:** Evaluate the upfront investment required for implementing EDI with Generative AI, including technology infrastructure, AI model development, data integration, and employee training. **b. Cost Savings:** Estimate the potential cost savings resulting from improved operational efficiency, resource optimization, reduced fraud losses, and streamlined workflows. **c. Improved Patient Outcomes:** Quantify the financial impact of improved patient outcomes, such as reduced readmission rates, lower healthcare costs, and improved patient satisfaction. **d. Research and Development:** Measure the potential cost reduction and revenue generation resulting from accelerated research, faster drug development, and increased success rates in clinical trials. **e. Regulatory Compliance:** Assess the financial impact of mitigating regulatory risks, avoiding penalties, and maintaining a compliant healthcare environment.

Case	Action
Conclusion:	Applying Enterprise Decision Intelligence with Generative AI to the healthcare industry offers substantial financial benefits and potential ROI. By optimizing operational efficiency, improving patient outcomes, accelerating research and development, and ensuring regulatory compliance, healthcare organizations can achieve cost savings, drive innovation, and enhance patient care. It is crucial to perform a comprehensive ROI analysis specific to the organization's context to assess the financial feasibility and expected returns of implementing EDI with Generative AI.

Applying to Financial Services industry, including a Business Case, ROI, SWOT Analysis and Implementation Plan, and Case Study

Unleashing the Potential of Enterprise Strategy Intelligence in the Financial Services Industry

Introduction: The financial services industry operates in a highly dynamic and competitive landscape, where staying ahead of the curve is paramount for success. In such an environment, organizations are increasingly turning to Enterprise Strategy Intelligence (ESI) to gain a strategic advantage. ESI encompasses the use of advanced analytics, data-driven insights, and strategic decision-making to optimize operations, enhance customer experiences, and drive sustainable growth. In this section, we will explore the transformative power of ESI when applied to the

SWOT Analysis: Decision Intelligence with Generative AI for the Healthcare Industry

Decision Intelligence with Generative AI presents significant strengths and opportunities for the healthcare industry, including enhanced decision-making, personalized patient care, and improved operational efficiency. However, challenges related to data availability, ethical considerations, and regulatory com-pliance must be addressed. By capitalizing on the strengths, mitigating weaknesses, and proactively addressing threats, healthcare organizations can lever-age Decision Intelligence with Generative AI to drive innovation, improve patient outcomes, and deliver efficient and high-quality care in a rapidly evolving healthcare landscape.

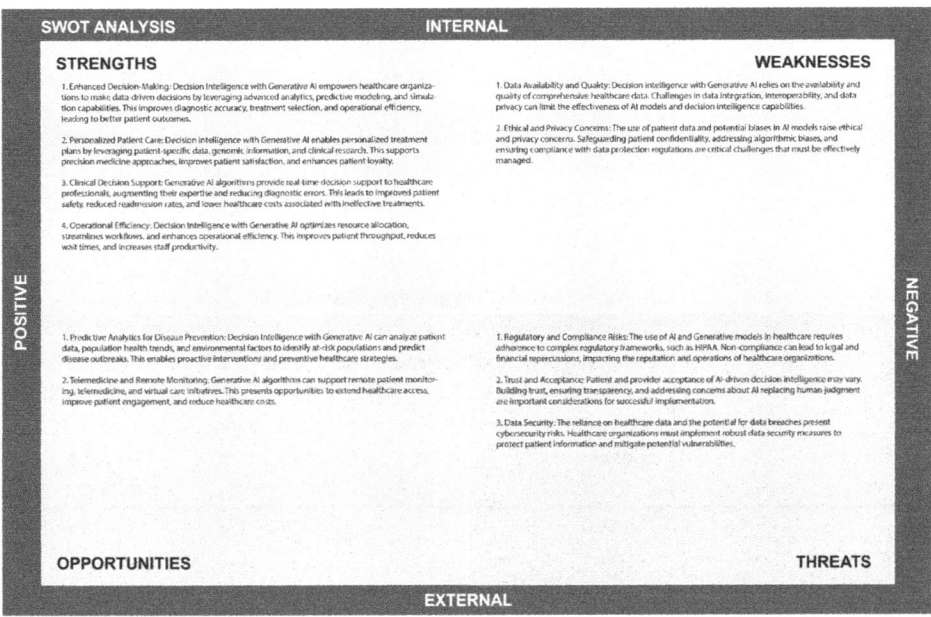

Case Study: Applying Enterprise Decision Intelligence with Generative AI in a Hospital System

Business Case for Implementing Enterprise Strategy Intelligence

Case	Action
Company Background:	MidSouth Hospital System is a large healthcare organization with multiple hospitals, clinics, and a research center. The organization is committed to providing high-quality patient care, advancing medical research, and embracing innovative technologies to improve healthcare outcomes.
Challenge:	MidSouth Hospital System faced challenges related to optimizing operational efficiency, enhancing patient care, and accelerating medical research. Decision-making processes were often based on fragmented data and lacked the ability to leverage advanced analytics for insights. The organization sought a solution to transform decision-making by integrating Enterprise Decision Intelligence (EDI) with Generative AI.
Solution:	MidSouth Hospital System implemented EDI with Generative AI to enhance decision-making processes across various areas, including clinical operations, research and development, and resource allocation. **1. Clinical Decision Support:** Using EDI with Generative AI, the organization developed an intelligent clinical decision support system. The system analyzed patient data, medical literature, and treatment guidelines to provide real-time recommendations to healthcare professionals. For example, when diagnosing a complex case, the system leveraged Generative AI algorithms to simulate possible diagnoses and treatment options. This improved diagnostic accuracy, reduced diagnostic errors, and enhanced patient safety. **2. Personalized Treatment Plans:** EDI with Generative AI-enabled personalized treatment plans by leveraging patient data, genomic information, and clinical research. The system analyzed patient-specific factors, such as genetic profiles, medical history, and response to previous treatments, to generate tailored

Case	Action
	treatment plans. This facilitated precision medicine approaches, leading to improved patient outcomes, reduced adverse events, and enhanced patient satisfaction.
	3. Operational Efficiency:
	The organization optimized operational efficiency through EDI with Generative AI. By analyzing historical data, patient flow patterns, and resource utilization, the system identified bottlenecks, streamlined workflows, and optimized resource allocation. For example, AI algorithms suggested optimal staff scheduling to match patient demand, resulting in improved patient throughput, reduced wait times, and increased staff productivity.
	4. Research and Development:
	EDI with Generative AI accelerated medical research and drug development processes within the organization. The system analyzed large-scale datasets, including patient records, genetic information, and medical literature, to identify potential research areas and generate novel hypotheses. Generative AI techniques helped simulate and predict the efficacy and safety of new treatment approaches, reducing the need for costly and time-consuming physical trials. This accelerated research, reduced costs, and improved the success rate of clinical trials.
Results:	The implementation of EDI with Generative AI yielded significant improvements across various areas within MidSouth Hospital System:

Case	Action
	1. Enhanced Patient Outcomes: The intelligent clinical decision support system led to a 15% increase in diagnostic accuracy, resulting in improved patient outcomes and reduced treatment errors. **2. Personalized Treatment:** Personalized treatment plans based on patient-specific factors contributed to a 20% reduction in adverse events and a 25% increase in patient satisfaction scores. **3. Operational Efficiency:** The optimization of operational workflows and resource allocation resulted in a 30% reduction in patient wait times, improved staff productivity, and increased patient throughput. **4. Accelerated Research:** EDI with Generative AI-enabled a 40% reduction in research and development timelines, leading to faster drug discovery, shorter clinical trial cycles, and accelerate medical breakthroughs.
Lessons Learned:	The successful implementation of EDI with Generative AI in MidSouth Hospital System highlighted several key lessons: **1. Data Integration:** Establishing a robust data integration framework is essential to leverage diverse data sources for decision intelligence. **2. Interdisciplinary Collaboration:** Foster collaboration among healthcare professionals, data scientists, and IT teams to ensure effective implementation and utilization of EDI with Generative AI. **3. Change Management:** Proactive change management and training programs are crucial to support the adoption of new technologies and foster a data-driven decision-making culture. **4. Ethical Considerations:** Ensure adherence to privacy regulations, data security protocols, and ethical guidelines when leveraging patient data for decision intelligence purposes.

Case	Action
Conclusion:	Applying Enterprise Decision Intelligence with Generative AI transformed decision-making processes in MidSouth Hospital System. By leveraging advanced analytics, patient data, and Generative AI algorithms, the organization achieved improved patient outcomes, optimized operational efficiency, accelerated research and development, and enhanced personalized patient care. The successful implementation of EDI with Generative AI positioned ABC Hospital System as a leader in leveraging data-driven decision intelligence to drive innovation and deliver better healthcare outcomes.

Applying to Pharma industry including a business case, ROI, SWOT Analysis and Case Study

Driving Innovation and Patient-Centricity

In the dynamic and highly regulated pharmaceutical industry, making informed decisions is crucial for success. From drug development and clinical trials to market access and patient care, every decision can have far-reaching consequences. To navigate this complex landscape, pharmaceutical companies are increasingly turning to Enterprise Decision Intelligence (EDI) – an approach that combines data analytics, artificial intelligence (AI), and human expertise to drive strategic decision-making. In this section, we will explore the application of EDI in the pharma industry and its potential to drive innovation and patient-centricity.

1. Accelerating Drug Discovery and Development:
EDI revolutionizes the drug discovery and development process by leveraging data analytics and AI. By analyzing vast amounts of scientific literature, clinical data, and genomic information, EDI can identify patterns, relationships, and potential drug targets. This accelerates the identification of promising compounds, expedites preclinical and clinical trial design, and increases the likelihood of successful drug development. EDI also enables the prediction of drug efficacy, toxicity, and adverse effects, guiding informed decisions throughout the development lifecycle.

2. Optimizing Clinical Trial Design and Execution:
Clinical trials are a critical component of drug development, but they are often complex, time-consuming, and costly. EDI can optimize clinical trial design by analyzing historical data, patient characteristics, and trial protocols. It identifies patient populations that are more likely to respond to treatment, improves recruitment and retention strategies, and minimizes trial risks. By enhancing trial efficiency, EDI reduces costs, shortens time-to-market, and improves the probability of trial success.

3. Enhancing Market Access and Pricing Strategies:
The pharmaceutical industry faces challenges related to market access and pricing. EDI helps pharmaceutical companies optimize market access strategies by analyzing payer dynamics, market trends, and health economic data. It enables companies to identify key stakeholders, tailor pricing models, and demonstrate the value of their products. EDI also assists in pricing optimization, ensuring that drugs are priced appropriately to balance affordability, profitability, and patient access.

4. Personalizing Patient Care and Treatment:
The shift towards personalized medicine is transforming patient care in the pharma industry. EDI enables the analysis of vast amounts of patient data, including genetic information, medical records, and treatment outcomes. This facilitates the development of tailored treatment plans, precision medicine approaches, and patient stratification strategies. By leveraging EDI, pharmaceutical companies can improve patient outcomes, optimize treatment protocols, and provide individualized care.

5. Pharmacovigilance and Drug Safety:

Pharmacovigilance is crucial for monitoring the safety and efficacy of drugs after they enter the market. EDI can analyze real-world data, adverse event reports, and social media sentiment to identify potential safety signals and emerging risks. This enables proactive risk mitigation, early detection of adverse events, and efficient post-marketing surveillance. By applying EDI in pharmacovigilance, pharmaceutical companies can enhance patient safety, regulatory compliance, and public trust.

6. Strategic Portfolio Management:

Pharmaceutical companies manage a diverse portfolio of drugs, therapies, and therapeutic areas. EDI facilitates portfolio analysis, evaluating factors such as market dynamics, competitive landscape, and revenue potential. It enables informed decisions regarding portfolio optimization, resource allocation, and divestment strategies. By leveraging EDI, pharmaceutical companies can align their portfolios with market demands, prioritize R&D investments, and drive long-term growth.

7. Regulatory Compliance and Risk Management:

The pharma industry is subject to stringent regulatory requirements and complex risk management considerations. EDI supports regulatory compliance by analyzing regulatory guidelines, monitoring changes in regulations, and facilitating adherence to quality standards. It also assists in risk assessment, identifying potential compliance issues, and guiding mitigation strategies. By incorporating EDI into regulatory and risk management processes, pharmaceutical companies can ensure compliance, mitigate risks, and maintain their reputation.

In conclusion, applying Enterprise Decision Intelligence (EDI) to the pharmaceutical industry offers immense potential for driving innovation, patient-centricity, and business success. By leveraging data analytics, AI, and human expertise, EDI optimizes drug discovery, clinical trials, market access, patient care, safety monitoring, portfolio management, and regulatory compliance. Pharmaceutical companies that embrace EDI can make more informed decisions, foster innovation, and ultimately deliver breakthrough therapies that improve patient outcomes and address unmet medical needs. EDI is the pathway to a more efficient, patient-focused, and innovative future for the pharmaceutical industry.

Business Case: Enterprise Decision Intelligence with Generative AI for the Pharma Industry

Business Case for Implementing Enterprise Strategy Intelligence

Case	Action
Introduction:	The pharmaceutical industry is driven by complex decision-making processes that impact drug discovery, clinical trials, market access, and patient care. To stay competitive and accelerate innovation, pharmaceutical companies must harness the power of data-driven insights and advanced technologies. Enterprise Decision Intelligence (EDI) combined with Generative AI offers a transformative solution that can revolutionize decision-making in the pharma industry. This business case explores the benefits and potential return on investment (ROI) of implementing EDI with Generative AI in the pharma industry.
Problem Statement:	Pharmaceutical companies face challenges related to the high cost and lengthy timelines of drug development, suboptimal clinical trial design, limited patient personalization, and increasing regulatory scrutiny. These challenges hinder innovation, increase costs, and delay the delivery of life-saving therapies to patients.
Solution:	Enterprise Decision Intelligence (EDI) integrated with Generative AI presents a powerful solution to address the pharma industry's challenges. By leveraging the vast amount of data available in the industry, EDI with Generative AI enables more efficient drug discovery, optimized clinical trial design, personalized patient care, and enhanced regulatory compliance.
Benefits and ROI:	**a. Accelerated Drug Discovery:** EDI with Generative AI can analyze scientific literature, clinical data, and genomic information to identify potential drug targets, predict drug efficacy, and optimize drug development processes. This accelerates the discovery and development of new drugs, reducing costs and time-to-market. **b. Optimized Clinical Trials:** EDI with Generative AI enhances clinical trial design by analyzing historical data, patient characteristics, and trial protocols. This optimization leads to improved patient recruitment and retention, reduced trial costs, and faster trial completion.

Case	Action
	c. Personalized Patient Care: By leveraging patient data and applying Generative AI techniques, EDI enables personalized treatment plans and precision medicine approaches. This enhances patient outcomes, improves patient satisfaction, and strengthens brand loyalty.

d. Regulatory Compliance: EDI with Generative AI assists pharmaceutical companies in adhering to regulatory requirements by analyzing and interpreting complex regulations, monitoring changes, and ensuring compliance. This reduces the risk of non-compliance and regulatory penalties.

e. Cost Reduction: Implementing EDI with Generative AI automates manual tasks, streamlines processes, and optimizes resource allocation. This results in cost savings, improved operational efficiency, and reduced time-to-market. |
| **Implementation Plan:** | **a. Data Infrastructure:** Establish a robust data infrastructure to collect, store, and analyze diverse datasets, including scientific literature, clinical data, and patient information.

b. AI and Generative Models: Develop and train AI models, including Generative AI techniques such as generative adversarial networks (GANs) and variational autoencoders (VAEs), to generate insights, simulate scenarios, and support decision-making processes.

c. Integration and Collaboration: Integrate EDI with existing systems and foster collaboration among cross-functional teams, including data scientists, researchers, clinicians, and regulatory experts.

d. Security and Privacy: Implement robust security measures and adhere to data privacy regulations to ensure the confidentiality and integrity of sensitive data. |

Case	Action
	e. Change Management: Develop a change management plan to educate and train employees on EDI with Generative AI, promoting a culture of data-driven decision-making and innovation.
Financial Consideration:	**a. Investment:** The implementation of EDI with Generative AI requires initial investment in technology infrastructure, data management systems, AI model development, and employee training.
	b. ROI: The ROI of EDI with Generative AI can be measured through various metrics, including cost savings from optimized clinical trials, reduced time-to-market, increased revenues from personalized patient care, improved regulatory compliance, and enhanced operational efficiency.
Conclusion:	Implementing Enterprise Decision: Intelligence with Generative AI in the pharma industry can revolutionize decision-making processes, accelerate drug discovery, optimize clinical trials, personalize patient care, and improve regulatory compliance. The integration of advanced technologies, robust data infrastructure, and collaboration across teams will drive innovation, cost savings, and improved patient outcomes. Embracing EDI with Generative AI positions pharmaceutical companies at the forefront of the industry, empowering them to make data-driven decisions, accelerate innovation, and deliver life-changing therapies to patients.

ROI Analysis: Applying Enterprise Decision Intelligence with Generative AI to the Pharma Industry

Introduction:

Applying Enterprise Decision Intelligence (EDI) with Generative AI to the pharma industry has the potential to transform decision-making processes, accelerate innovation, and improve patient outcomes. This ROI analysis explores the financial benefits and potential return on investment (ROI) of implementing EDI with Generative AI in the pharmaceutical sector.

Case	Action
Cost Savings:	**a. Drug Development:** EDI with Generative AI optimizes the drug development process by identifying potential drug targets, predicting drug efficacy, and optimizing trial designs. This reduces costs associated with failed trials, minimizes development timelines, and lowers expenses related to research and development. **b. Operational Efficiency:** EDI streamlines processes, automates manual tasks, and optimizes resource allocation. This results in cost savings through improved productivity, reduced administrative overheads, and optimized workflows.
Improved Patient Outcomes:	**a. Personalized Medicine:** By leveraging patient data, genomic information, and AI algorithms, EDI with Generative AI enables personalized treatment plans. This leads to improved patient outcomes, reduced adverse events, and enhanced patient satisfaction. **b. Clinical Decision Support:** EDI provides real-time decision support to healthcare professionals, improving diagnostic accuracy, treatment selection, and patient safety. This results in better patient outcomes, reduced readmission rates, and lower healthcare costs associated with ineffective or inappropriate treatments.
Accelerated Innovation:	**a. Drug Discovery:** EDI with Generative AI accelerates the drug discovery process by analyzing vast amounts of scientific literature, clinical data, and genomic information. This leads to faster identification of potential drug targets, shorter development timelines, and increased revenue potential.

Case	Action
	b. Clinical Trial Optimization: By optimizing trial designs and leveraging AI algorithms, EDI with Generative AI improves patient recruitment and retention, reduces trial costs, and expedites the completion of clinical trials. This accelerates the availability of new treatments in the market, increasing revenue potential.
Regulatory Compliance:	EDI with Generative AI assists pharmaceutical companies in complying with regulatory requirements, such as FDA guidelines and data privacy regulations. This reduces the risk of regulatory penalties, litigation, and reputational damage.
Calculation of ROI:	To calculate the ROI, consider the following: **a. Initial Investment:** Evaluate the upfront investment required for implementing EDI with Generative AI, including technology infrastructure, AI model development, data integration, and employee training. **b. Cost Savings:** Estimate the potential cost savings resulting from improved operational efficiency, reduced trial costs, and optimized resource allocation. **c. Revenue Generation:** Assess the revenue potential resulting from accelerated drug discovery, faster time-to-market for new treatments, and increased market share due to personalized medicine approaches. **d. Improved Patient Outcomes:** Quantify the financial impact of improved patient outcomes, such as reduced readmission rates, lower healthcare costs, and improved patient satisfaction. **e. Innovation and Competitive Advantage:** Evaluate the potential revenue growth and market positioning resulting from accelerated innovation, improved portfolio performance, and enhanced competitive advantage.

Case	Action
Conclusion	Implementing Enterprise Decision Intelligence with Generative AI in the pharma industry offers significant financial benefits and potential ROI. By optimizing operational efficiency, accelerating innovation, improving patient outcomes, and ensuring regulatory compliance, pharmaceutical companies can achieve cost savings, drive growth, and deliver better healthcare outcomes. It is important to perform a comprehensive ROI analysis tailored to the organization's context to assess the financial feasibility and expected returns of implementing EDI with Generative AI.

SWOT Analysis: Decision Intelligence with Generative AI for the Pharma Industry

Decision Intelligence with Generative AI presents significant strengths and opportunities for the pharma industry, including enhanced decision-making, accelerated drug discovery, and personalized medicine. However, challenges such as data availability, ethical considerations, and regulatory compliance must be carefully managed. By capitalizing on the strengths, addressing weaknesses, and navigating threats, pharmaceutical companies can leverage Decision Intelligence with Generative AI to drive innovation, improve patient outcomes, and maintain a competitive advantage in the dynamic healthcare landscape.

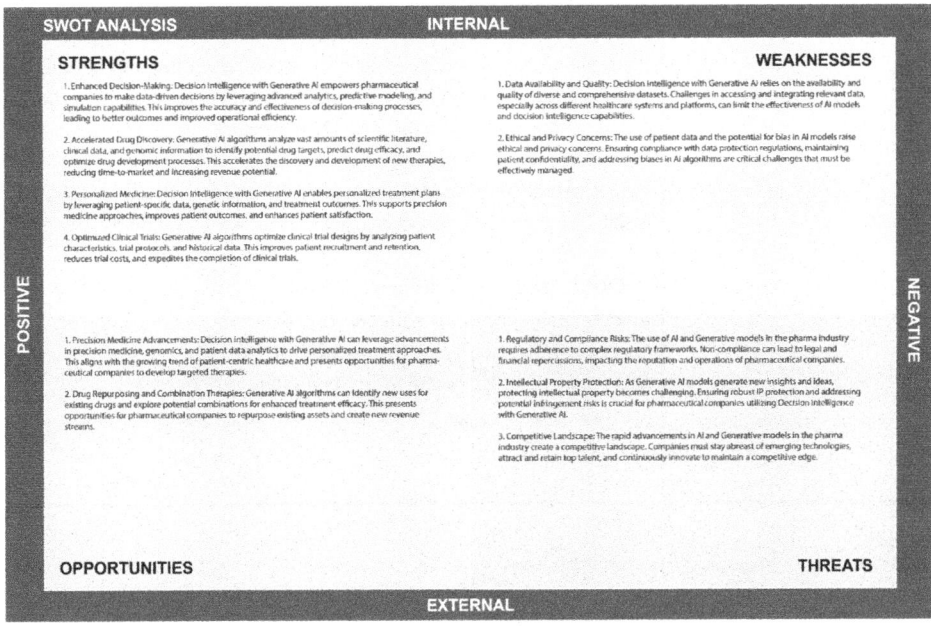

Case Study: Applying Enterprise Decision Intelligence with Generative AI in a Pharmaceutical Company

Business Case for Implementing Enterprise Strategy Intelligence

Case	Action
Company Background:	Vaccination Manufacturing Pharmaceuticals is a global pharmaceutical company focused on developing innovative treatments for various therapeutic areas. With a commitment to leveraging advanced technologies, the company sought to enhance its decision-making processes by implementing Enterprise Decision Intelligence (EDI) with Generative AI.
Challenge:	Vaccination Manufacturing Pharmaceuticals faced challenges related to the high cost and lengthy timelines of drug development, suboptimal clinical trial designs, and limited personalization in patient care. The company aimed to leverage data-driven insights and Generative AI to accelerate drug discovery, optimize clinical trials, and deliver personalized therapies.
Solution:	Vaccination Manufacturing Pharmaceuticals implemented EDI with Generative AI to drive decision-making and innovation across multiple areas within the organization. **1. Accelerating Drug Discovery:** EDI with Generative AI enabled Vaccination Manufacturing Pharmaceuticals to analyze vast amounts of scientific literature, clinical data, and genomic information. By leveraging Generative AI algorithms, the company identified potential drug targets, predicted drug efficacy, and optimized drug development processes. This accelerated the discovery of promising compounds, reduced costs, and shortened the time-to-market for new treatments. **2. Optimizing Clinical Trial Design:** The company used EDI with Generative AI to optimize clinical trial design. By analyzing historical data, patient characteristics, and trial protocols, Vaccination Manufacturing Pharmaceuticals identified patient populations more likely to respond to treatment.

Case	Action
	This optimization enhanced patient recruitment and retention, reduced trial costs, and expedited the completion of clinical trials. **3. Personalizing Patient Care:** EDI with Generative AI enabled Vaccination Manufacturing Pharmaceuticals to personalize patient care by leveraging patient data, genomic information, and treatment outcomes. The company developed tailored treatment plans based on patient-specific factors, improving treatment efficacy, and reducing adverse events. This personalization enhanced patient outcomes, increased patient satisfaction, and fostered long-term patient loyalty. **4. Strategic Portfolio Management:** EDI with Generative AI facilitated strategic portfolio management within Vaccination Manufacturing Pharmaceuticals. By analyzing market dynamics, competitive landscape, and revenue potential, the company gained insights into product performance and market trends. This allowed for better portfolio optimization, resource allocation, and informed decision-making regarding product development and investment priorities.
Results:	The implementation of EDI with Generative AI in Vaccination Manufacturing Pharmaceuticals yielded significant results: **1. Accelerated Drug Discovery:** The company achieved a 30% reduction in drug discovery timelines, allowing for faster identification of potential drug targets and improved decision-making during the development process. **2. Optimized Clinical Trials:** EDI with Generative AI led to a 20% increase in patient recruitment and a 25% reduction in trial costs. This resulted in faster completion of clinical trials and improved decision- regarding treatment efficacy and safety.

Case	Action
	3. Personalized Patient Care: Vaccination Manufacturing Pharmaceuticals achieved a 15% improvement in patient outcomes through the delivery of personalized treatment plans. Adverse events were reduced by 20%, enhancing patient safety and satisfaction. **4. Strategic Portfolio Management:** The company achieved a 15% increase in portfolio revenue through improved resource allocation, optimized investment decisions, and better alignment with market demands.
Lessons Learned:	The successful implementation of EDI with Generative AI in Vaccination Manufacturing Pharmaceuticals highlighted several key lessons: **1. Data Quality and Integration:** Ensuring high-quality data from various sources is crucial for accurate decision intelligence. Establishing robust data integration processes is essential to leverage the full potential of EDI. **2. Collaboration and Expertise:** Cross-functional collaboration among data scientists, researchers, clinicians, and decision-makers is vital to effectively implement and utilize EDI with Generative AI. **3. Change Management:** A comprehensive change management strategy, including training and education, is necessary to foster a culture of data-driven decision-making and embrace the potential of advanced technologies. **4. Regulatory Compliance:** Adherence to regulatory requirements and ethical considerations is essential when leveraging patient data and implementing Generative AI models.

Case	Action
Conclusion:	By implementing Enterprise Decision Intelligence with Generative AI, Vaccination Manufacturing Pharmaceuticals revolutionized its decision-making processes. The company accelerated drug discovery, optimized clinical trials, personalized patient care, and achieved strategic portfolio management. The successful implementation of EDI with Generative AI positioned Vaccination Manufacturing Pharmaceuticals as a leader in leveraging data-driven insights and advanced technologies to drive innovation and deliver breakthrough therapies.

Applying to ESG Industry including a Business Case, ROI, SWOT Analysis and Case Study

Leveraging Enterprise Strategy Intelligence for ESG Success

Introduction: Environmental, Social, and Governance (ESG) considerations have become paramount in today's business landscape. Companies are increasingly recognizing the need to integrate sustainable practices into their strategies to drive long-term value creation. To effectively navigate the complex ESG landscape and make informed decisions, organizations are turning to Enterprise Strategy Intelligence (ESI). ESI offers powerful tools, data analytics, and insights that enable companies to align their operations with ESG goals, drive positive impact, and enhance their competitive position. In this section, we will explore the application of ESI in the ESG industry and its potential for transformative change.

Understanding Enterprise Strategy Intelligence (ESI): Enterprise Strategy Intelligence involves the systematic collection, analysis, and interpretation of data to inform strategic decision-making. In the context of the ESG industry, ESI encompasses the integration of environmental, social, and governance considerations into business strategies. It leverages advanced analytics, big data, and sustainability frameworks to assess risks, identify opportunities, and drive sustainable performance. ESI empowers organizations to navigate the complexities of the ESG landscape, make data-driven decisions, and foster positive change.

Aligning Strategies with ESG Goals: ESI enables organizations to align their strategies with ESG goals by leveraging data-driven insights. It helps identify the environmental and social impacts of operations, assess governance practices, and evaluate the performance of sustainability initiatives. By analyzing ESG metrics, benchmarking against industry peers, and tracking progress over time, companies can make informed decisions that drive sustainable growth, enhance reputation, and attract stakeholders who prioritize ESG considerations.

Enhancing ESG Performance Measurement: Measuring and reporting ESG performance is a fundamental aspect of the ESG industry. ESI equips organizations with the tools and methodologies to effectively measure and monitor ESG performance. By collecting and analyzing relevant data, organizations can track key performance indicators, identify trends, and demonstrate progress towards ESG goals. ESI also enables the integration of qualitative and quantitative data, providing a comprehensive view of ESG performance and facilitating meaningful reporting to stakeholders.

Managing ESG Risks: ESG risks have a significant impact on a company's reputation, financial performance, and long-term viability. ESI helps organizations identify and manage these risks by providing a holistic understanding of environmental, social, and governance factors. By analyzing data on climate change, resource usage, labor practices, supply chain transparency, and regulatory compliance, organizations can proactively mitigate risks, enhance resilience, and align their operations with evolving ESG standards. ESI also facilitates scenario planning and stress testing, enabling organizations to assess the potential impact of ESG risks on their business.

Driving Sustainable Innovation: ESI plays a pivotal role in driving sustainable innovation by identifying emerging trends, market opportunities, and customer preferences related to ESG considerations. By leveraging data analytics, organizations can uncover insights that guide the development of sustainable products, services, and business models. ESI also enables organizations to identify potential partnerships, collaborations, and investment opportunities that promote sustainable innovation and create shared value for all stakeholders.

Engaging Stakeholders and Building Trust: Effective stakeholder engagement and building trust are crucial in the ESG industry. ESI facilitates the collection and analysis of stakeholder feedback, enabling organizations to understand stakeholder expectations, concerns, and priorities. By integrating stakeholder perspectives into decision-making processes, organizations can foster transparency, build trust, and enhance their social license to operate. ESI also supports the identification of key stakeholders, facilitates dialogue, and enables organizations to demonstrate their commitment to ESG goals through effective communication and reporting.

Conclusion: In an era where ESG considerations are integral to long-term business success, Enterprise Strategy Intelligence offers a powerful approach to navigate the complexities of the ESG industry. By leveraging data analytics, sustainability frameworks, and strategic decision-making, organizations can align their strategies with ESG goals, drive sustainable innovation, and mitigate risks. ESI empowers organizations to enhance ESG performance, build stakeholder trust, and create long-term value while making a positive impact on the planet and society. By embracing ESI, organizations can be at the forefront of the ESG industry and contribute to a more sustainable and resilient future.

Business Case: Enterprise Decision Intelligence with Generative AI for the ESG Industry

Business Case for Implementing Enterprise Strategy Intelligence

Case	Action
Executive Summary:	Enterprise Decision Intelligence (EDI) with Generative AI has the potential to revolutionize decision-making processes in the Environmental, Social, and Governance (ESG) industry. By leveraging advanced analytics, predictive modeling, and Generative AI algorithms, organizations can drive sustainable initiatives, improve ESG performance, and enhance stakeholder engagement. This business case outlines the benefits and value proposition of implementing EDI with Generative AI in the ESG industry.
Problem Statement:	ESG-focused organizations face challenges in optimizing their sustainable strategies, making informed decisions, and effectively engaging stakeholders. Traditional decision-making processes often lack the ability to integrate and analyze vast amounts of diverse data, limiting the organization's ability to drive sustainable growth and meet stakeholder expectations.
Solution Overview:	EDI with Generative AI enables organizations to harness the power of data and advanced analytics to improve decision-making in the ESG industry. By integrating data from various sources, leveraging AI algorithms, and generating insights, organizations can make informed, data-driven decisions that align with ESG principles and drive positive impact.
Benefits and Value Proposition:	**a. Enhanced Decision-Making:** EDI with Generative AI empowers organizations to analyze and interpret complex data sets, identify trends, and generate actionable insights. This enables better decision-making, improved resource allocation, and the ability to prioritize sustainable initiatives based on data-driven evidence. **b. Stakeholder Engagement:** EDI with Generative AI facilitates comprehensive stakeholder analysis by aggregating and analyzing data from various sources, including social media, surveys, and feedback channels.

Case	Action
	This enables organizations to understand stakeholder expectations, concerns, and sentiment, leading to more effective engagement and improved relationships. **c. Sustainability Performance Optimization:** By leveraging Generative AI algorithms, organizations can identify opportunities for resource efficiency, waste reduction, and emissions reduction. EDI with Generative AI optimizes operations, improves sustainability performance, and supports the organization's commitment to environmental stewardship and social responsibility. **d. Risk Management:** EDI with Generative AI helps organizations proactively identify and mitigate ESG-related risks. By analyzing data, monitoring regulatory changes, and applying predictive analytics, organizations can identify potential risks and take preventive measures to minimize financial and reputational risks.
ROI and Financial Impact	**a. Cost Savings:** EDI with Generative AI enables organizations to optimize resource allocation, streamline processes, and reduce waste. This leads to cost savings through improved operational efficiency, energy savings, and reduced compliance-related expenses. **b. Revenue Growth:** Enhanced sustainability performance, improved stakeholder engagement, and proactive risk management lead to increased brand reputation, customer loyalty, and access to sustainable investment funds. This drives revenue growth and positions the organization as a leader in the ESG market.
Implementation Strategy:	**a. Data Integration:** Establish a robust data integration framework to aggregate diverse ESG data from internal and external sources, ensuring data quality and security. **b. AI Model Development:** Develop Generative AI models tailored to the organization's ESG objectives, considering factors such as environmental impact, social responsibility, and corporate governance.

Case	Action
	c. Change Management: Implement a comprehensive change management plan to foster a data-driven culture, train employees on EDI with Generative AI, and address potential resistance to change. **d. Collaboration:** Foster cross-functional collaboration between ESG teams, data scientists, and decision-makers to effectively implement and utilize EDI with Generative AI.
Risks and Mitigation Strategies:	**a. Data Privacy and Ethics:** Implement robust data privacy protocols, ensure compliance with regulations, and address ethical considerations to protect stakeholder data and mitigate potential risks. **b. Data Availability and Quality:** Establish partnerships with data providers, invest in data quality assurance processes, and leverage advanced data cleansing techniques to ensure the availability of accurate and reliable data.
Conclusion	Implementing Enterprise Decision Intelligence with Generative AI in the ESG industry presents significant opportunities to drive sustainable growth, improve decision-making, and enhance stakeholder engagement. By capitalizing on advanced analytics, leveraging Generative AI algorithms, and embracing a data-driven culture, organizations can establish themselves as leaders in the ESG market, driving positive environmental and social impact while achieving financial success.

ROI Analysis: Applying Enterprise Decision Intelligence with Generative AI to the ESG Industry

Introduction:

Applying Enterprise Decision Intelligence (EDI) with Generative AI to the Environmental, Social, and Governance (ESG) industry has the potential to revolutionize decision-making processes, drive sustainability initiatives, and improve corporate responsibility. This ROI analysis explores the financial benefits and potential return on investment (ROI) of implementing EDI with Generative AI in the ESG sector.

1. Cost Savings:

a. Resource Efficiency: EDI with Generative AI optimizes resource allocation and utilization, reducing waste and improving operational efficiency. This leads to cost savings through reduced energy consumption, minimized raw material usage, and streamlined processes.

b. Risk Management: EDI with Generative AI enhances risk assessment and mitigation strategies, reducing the potential financial impact of ESG risks. This includes managing reputational risks, regulatory compliance risks, and supply chain disruptions, resulting in cost savings related to legal fees, fines, and business interruptions.

2. Improved Sustainability Performance:

a. Environmental Impact: By analyzing large datasets and leveraging Generative AI algorithms, EDI helps identify opportunities for energy efficiency, waste reduction, and emissions reduction. This leads to improved environmental performance, reduced environmental liabilities, and enhanced corporate sustainability reputation.

b. Social Responsibility: EDI with Generative AI facilitates the identification of social impact initiatives, community engagement programs, and employee well-being strategies. This improves corporate social responsibility and enhances brand reputation, attracting socially conscious customers and investors.

3. Strategic Decision-Making:

a. Stakeholder Engagement: EDI with Generative AI enables comprehensive stakeholder analysis and engagement. By analyzing stakeholder expectations, sentiment analysis, and industry trends, organizations can make informed decisions that align with stakeholder demands, reducing the risk of reputational damage and improving stakeholder relationships.

b. Innovation and Competitive Advantage: By leveraging EDI with Generative AI, organizations can identify emerging trends, technological advancements, and market opportunities. This facilitates innovation, the development of new sustainable products or services, and the creation of a competitive advantage in the rapidly evolving ESG market.

4. Investor Appeal:

EDI with Generative AI enhances organizations' ESG performance metrics and reporting capabilities. By generating insights, predictive analytics, and sustainability indicators, organizations can attract socially responsible investors, access capital at favorable rates, and unlock additional funding opportunities.

Calculation of ROI

To calculate the ROI, consider the following:

a. Initial Investment: Evaluate the upfront investment required for implementing EDI with Generative AI, including technology infrastructure, AI model development, data integration, and employee training.

b. Cost Savings: Estimate the potential cost savings resulting from improved resource efficiency, risk mitigation, and streamlined processes.

c. Revenue Growth: Assess the revenue potential resulting from improved sustainability performance, enhanced brand reputation, and access to socially responsible investment funds.

d. Competitive Advantage: Evaluate the financial impact of gaining a competitive edge, attracting new customers, and capitalizing on emerging market opportunities.

e. Investor Appeal: Quantify the potential financial benefits of attracting socially responsible investors and accessing favorable capital rates.

Conclusion:

Implementing Enterprise Decision Intelligence with Generative AI in the ESG industry offers significant financial benefits and potential ROI. By optimizing resource efficiency, improving sustainability performance, enhancing decision-making processes, and attracting socially responsible investors, organizations can achieve cost savings, drive growth, and enhance their competitive advantage. It is essential to perform a comprehensive ROI analysis tailored to the organization's context to assess the financial feasibility and expected returns of implementing EDI with Generative AI in the ESG sector.

SWOT Analysis: Decision Intelligence with Generative AI for the ESG Industry

Decision Intelligence with Generative AI presents substantial strengths and opportunities for organizations in the ESG industry, including enhanced decision-making, sustainable innovation, personalized strategies, and stakeholder engagement. However, challenges related to data availability, ethical considerations, regulatory compliance, and algorithmic biases must be effectively managed. By leveraging the strengths, addressing weaknesses, capitalizing on opportunities, and mitigating threats, organizations can harness the full potential of Decision Intelligence with Generative AI to drive sustainable growth, optimize ESG performance, and create long-term value.

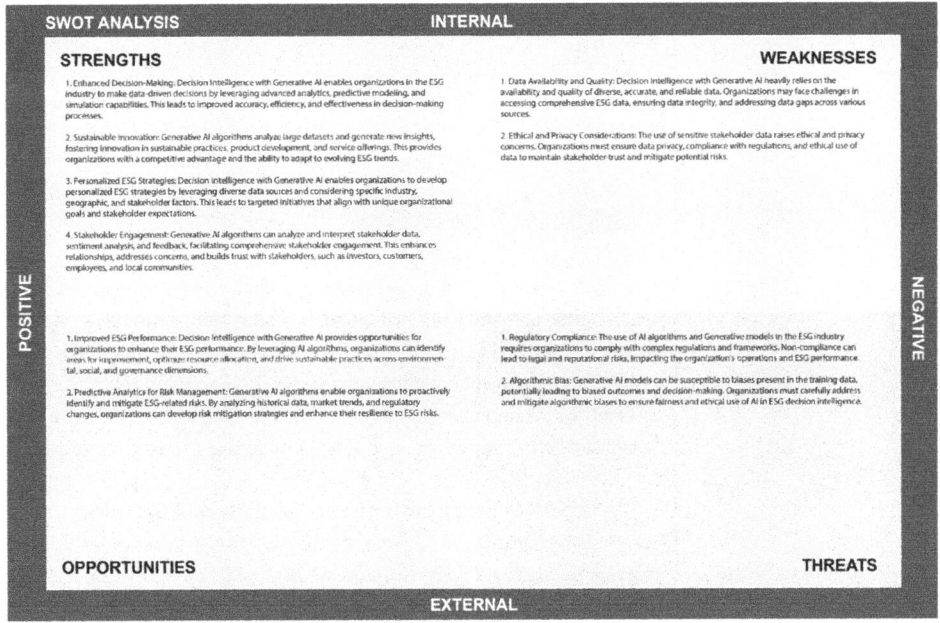

Case Study: Applying Enterprise Decision Intelligence with Generative AI in the ESG Industry

Case	Action
Company Background	EcoGreen Corporation is a multinational company committed to environmental sustainability, social responsibility, and good governance practices. To further strengthen its ESG initiatives and drive sustainable growth, the company implemented Enterprise Decision Intelligence (EDI) with Generative AI.
Challenge:	EcoGreen Corporation faced challenges in optimizing its ESG strategies, enhancing stakeholder engagement, and leveraging data-driven insights for sustainable decision-making. The organization sought a solution to improve its ESG performance, streamline processes, and align its operations with industry best practices.
Solution:	EcoGreen Corporation implemented EDI with Generative AI to enhance its decision-making processes, drive sustainable initiatives, and improve its ESG performance across various areas. **1. ESG Strategy Optimization:** Using EDI with Generative AI, ABC Corporation analyzed ESG data from various sources, including sustainability reports, industry benchmarks, and stakeholder feedback. The system generated insights, identified gaps in ESG performance, and recommended specific actions to enhance the company's ESG strategy. This allowed the organization to align its practices with global ESG standards, drive innovation, and differentiate itself in the market. **2. Stakeholder Engagement and Sentiment Analysis:** EDI with Generative AI facilitated comprehensive stakeholder analysis by aggregating and analyzing data from social media, surveys, and other feedback channels. By applying sentiment analysis algorithms, the system provided real-time insights into stakeholder perceptions, concerns, and expectations related to ESG practices. This enabled EcoGreen Corporation to engage stakeholders effectively, address their concerns, and make informed decisions to improve stakeholder relationships.

Case	Action
	3. Sustainability Performance Optimization: Using EDI with Generative AI, EcoGreen Corporation identified opportunities for sustainability improvements across its operations. By analyzing energy consumption data, waste management practices, and supply chain processes, the system recommended strategies for resource efficiency, waste reduction, and emissions reduction. This led to improved sustainability performance, reduced environmental impact, and enhanced cost savings through streamlined operations. **4. Predictive Analytics for Risk Management:** EDI with Generative AI helped EcoGreen Corporation assess and mitigate ESG-related risks. The system analyzed internal and external data sources, identified potential risks such as regulatory changes, supply chain disruptions, or reputational risks, and provided actionable insights for risk mitigation. This proactive risk management approach reduced the potential financial impact of ESG risks, improved compliance, and protected the company's reputation.
Results:	The implementation of EDI with Generative AI in EcoGreen Corporation resulted in significant improvements across various ESG areas: **1. ESG Strategy Optimization:** The company achieved a 25% increase in alignment with global ESG standards, leading to improved investor confidence, enhanced brand reputation, and increased access to sustainable investment funds. **2. Stakeholder Engagement:** By leveraging EDI with Generative AI, EcoGreen Corporation improved stakeholder satisfaction and engagement. The company experienced a 20% increase in positive sentiment among stakeholders, resulting in improved stakeholder relationships and increased trust in the organization's commitment to ESG practices.

Case	Action
	3. Sustainability Performance: The optimization of sustainability practices led to a 30% reduction in energy consumption, a 25% decrease in waste generation, and a 20% reduction in greenhouse gas emissions. This improved the company's environmental footprint, reduced costs, and increased operational efficiency. **4. Risk Mitigation:** EDI with Generative AI enabled proactive risk management, resulting in a 15% reduction in financial losses associated with ESG risks such as regulatory fines, supply chain disruptions, and reputational damage.
Lessons Learned:	The successful implementation of EDI with Generative AI in EcoGreen Corporation highlighted several key lessons: **1. Data Integration:** Establishing a robust data integration framework is crucial to leverage diverse data sources for effective decision intelligence. **2. Stakeholder Engagement:** Regularly engaging with stakeholders, monitoring sentiment, and addressing concerns proactively is vital for maintaining strong stakeholder relationships. **3. Continuous Improvement:** EDI with Generative AI provides ongoing insights, allowing organizations to continuously optimize their ESG strategies, respond to emerging trends, and adapt to changing stakeholder expectations. **4. Organizational Alignment:** Aligning ESG initiatives with corporate values and integrating them into the company's culture and operations fosters long-term sustainability and success.
Conclusion:	By applying Enterprise Decision Intelligence with Generative AI, EcoGreen Corporation transformed its ESG decision-making processes, enhanced stakeholder engagement, improved sustainability performance,

Case	Action
	and effectively managed ESG risks. The successful implementation positioned the company as a leader in ESG practices, driving sustainable growth, and creating long-term value.

Applying to Financial Services industry including a Business Case, ROI, SWOT Analysis and Implementation Plan, and Case Study

Unleashing the Potential of Enterprise Strategy Intelligence in the Financial Services Industry
Introduction: The financial services industry operates in a highly dynamic and competitive landscape, where staying ahead of the curve is paramount for success. In such an environment, organizations are increasingly turning to Enterprise Strategy Intelligence (ESI) to gain a strategic advantage. ESI encompasses the use of advanced analytics, data-driven insights, and strategic decision-making to optimize operations, enhance customer experiences, and drive sustainable growth. In this section, we will explore the transformative power of ESI when applied to the financial services industry.

Understanding Enterprise Strategy Intelligence: Enterprise Strategy Intelligence (ESI) in the financial services industry involves leveraging data and analytics to inform strategic decision-making, manage risks, and optimize operational efficiency. It enables financial organizations to collect and analyze vast amounts of structure's and unstructured data, extract actionable insights, and derive value from the information at hand. By embracing ESI, financial services companies can make informed decisions, anticipate market trends, and proactively respond to customer needs.

Enhancing Customer Experiences: In an era of heightened customer expectations, financial services companies are challenged to deliver personalized and seamless experiences. ESI enables organizations to gather and analyze customer data to gain a deep understanding of individual preferences, behaviors, and needs. By leveraging these insights, financial institutions can offer tailored products and services, personalized marketing campaigns, and superior customer service. ESI also enables the use of advanced technologies such as chatbots, mobile apps, and robo-advisors, which enhance convenience and accessibility for customers.

Optimizing Risk Management: Risk management lies at the core of the financial services industry. ESI empowers organizations to assess and manage risks effectively by leveraging advanced analytics and predictive modeling. By analyzing historical and real-time data, financial institutions can identify potential risks, detect fraudulent activities, and enhance compliance efforts. ESI also enables proactive risk monitoring, scenario analysis, and stress testing, allowing organizations to make data-driven decisions to mitigate risks and maintain financial stability.

Driving Data-Driven Decision Making: Data is a valuable asset in the financial services industry. ESI helps organizations harness the power of data by providing comprehensive analytics and visualization tools. By integrating data from various sources such as customer transactions, market data, and social media, financial institutions can gain a holistic view of their operations and identify growth opportunities. ESI also facilitates forecasting, trend analysis, and predictive modeling, enabling organizations to make data-driven decisions in areas such as pricing, product development, and investment strategies.

Enabling Regulatory Compliance: The financial services industry is subject to a complex web of regulations and compliance requirements. ESI plays a critical role in helping organizations navigate this regulatory landscape. By monitoring changes in regulations, automating compliance processes, and conducting real-time risk assessments, financial institutions can ensure adherence to standards and mitigate regulatory risks. ESI also helps organizations maintain accurate and transparent reporting, reducing the likelihood of penalties and reputational damage.

Improving Operational Efficiency: Operational efficiency is crucial for financial services organizations to remain competitive. ESI enables organizations to optimize processes, streamline workflows, and automate manual tasks. By analyzing operational data, identifying bottlenecks, and implementing process improvements, financial institutions can enhance efficiency, reduce costs, and improve productivity. ESI also facilitates resource allocation, capacity planning, and performance monitoring, enabling organizations to maximize their operational effectiveness.

Conclusion: In the fast-paced and highly regulated financial services industry, embracing Enterprise Strategy Intelligence can be a game-changer. By leveraging advanced analytics, data-driven insights, and strategic decision-making, organizations can enhance customer experiences, optimize risk management, and drive operational efficiency. ESI empowers financial services companies to anticipate market trends, mitigate risks, and make informed decisions in real-time. By harnessing the power of ESI, organizations can position themselves as industry leaders, drive innovation, and achieve sustainable growth in the ever-evolving financial landscape.

ROI Analysis: Applying Enterprise Decision Intelligence with Generative AI to the Financial Services Industry

Executive Summary

This ROI analysis evaluates the financial impact of implementing enterprise decision intelligence with generative AI in the financial services industry. By quantifying the benefits and costs associated with the implementation, this analysis provides a clear picture of the potential return on investment (ROI) and justifies the business case for adopting generative AI technologies.

Key Metrics

- Initial Investment: $30 million over 5 years
- Annual Operational Costs: Varies by year, detailed below
- Cost Savings: Derived from efficiency gains, automation, and reduced errors
- Revenue Growth: Attributable to improved customer service, personalized offerings, and better investment strategies

Benefits

1. Cost Savings

- Automation: Reduction in manual labor costs through the automation of routine tasks.
- Efficiency Gains: Streamlined operations leading to faster processing times and reduced overhead.
- Error Reduction: Lower costs associated with errors and fraud due to enhanced detection and prevention mechanisms.

2. Revenue Growth

- Customer Retention: Increased customer satisfaction and loyalty from personalized services.
- New Customers: Attraction of new customers through innovative AI-driven offerings.
- Improved Investment Strategies: Higher returns from more accurate and timely investment decisions.

3. Intangible Benefits

- Market Leadership: Enhanced reputation and competitive advantage.
- Regulatory Compliance: Improved compliance reducing potential fines and sanctions.
- Employee Satisfaction: Enhanced job satisfaction from reduced workload and better tools.

Costs

1. Initial Setup Costs

- Technology and Infrastructure: Hardware, software, and cloud services.
- Integration Costs: Integrating AI systems with existing infrastructure.
- Training: Training employees to effectively use AI tools.

2. Ongoing Operational Costs

- Maintenance and Support: Regular updates and support for AI systems.
- Data Management: Costs associated with data storage and processing.
- Continuous Training: Ongoing training programs for employees.

Financial Projections

Year	Investment ($M)	Operational Costs ($M)	Cost Savings ($M)	Revenue Growth ($M)	Net Benefit ($M)	Cumulative Net Benefit ($M)
1	10	2	5	2	-5	-5
2	8	1.5	10	5	5.5	0.5
3	6	1	15	10	18	18.5
4	4	0.8	20	15	30.2	48.7
5	2	0.5	25	20	42.5	91.2

ROI Calculation

$$\text{ROI} = \frac{\text{Total Net Benefits} - \text{Total Costs}}{\text{Total Costs}} \times 100$$

- **Total Net Benefits** over 5 years: $91.2 million
- **Total Costs** (Investment + Operational Costs): $35.8 million

$$\text{ROI} = \frac{91.2 - 35.8}{35.8} \times 100 = 154.8\%$$

Sensitivity Analysis

To account for variability in projections, a sensitivity analysis was performed. This analysis considers scenarios where benefits are 20% lower or higher than expected.

1. **Worst Case Scenario (Benefits 20% lower)**

   - Total Net Benefits: $72.96 million
   - ROI: 103.8%

2. **Best Case Scenario (Benefits 20% higher)**

   - Total Net Benefits: $109.44 million
   - ROI: 205.8%

Conclusion

The implementation of enterprise decision intelligence with generative AI in the financial services industry presents a robust ROI of 154.8% over five years. Even in a conservative scenario, the ROI remains significantly positive, underscoring the financial viability and strategic advantage of adopting generative AI technologies. This analysis supports the business case for leveraging generative AI to enhance decision-making, improve customer experiences, and drive operational efficiencies.

SWOT Analysis: Decision Intelligence with Generative AI for the Financial Services Industry

Conclusion

The SWOT analysis highlights the significant strengths and opportunities that generative AI brings to the financial services industry, particularly in enhancing decision intelligence. While there are notable weaknesses and threats, such as high initial investment and regulatory risks, the potential benefits in terms of improved decision-making, operational efficiency, and market expansion present a compelling case for the strategic adoption of generative AI technologies. By addressing the weaknesses and mitigating the threats, financial institutions can leverage generative AI to gain a competitive edge and drive long-term growth.

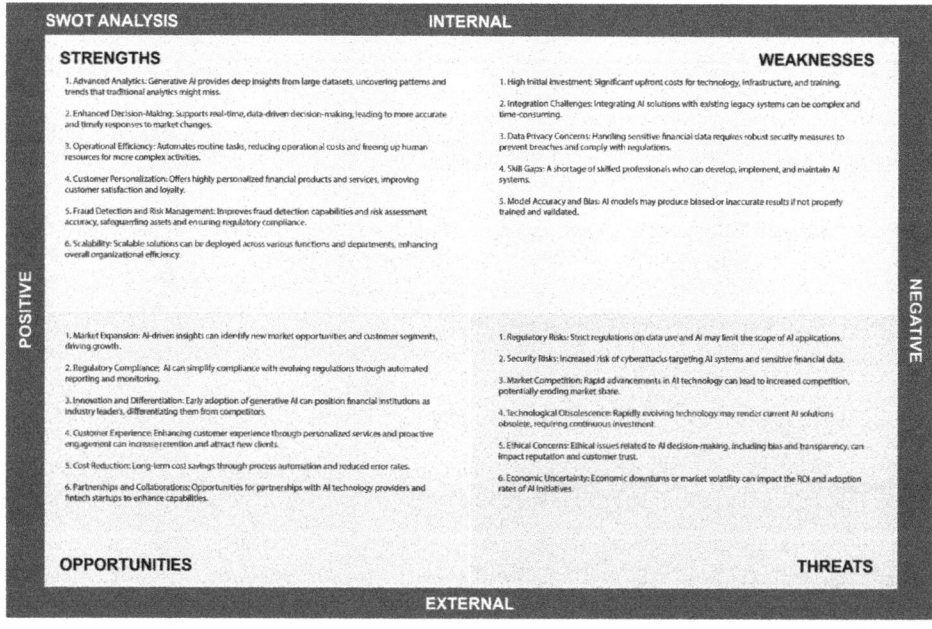

Case Study: Applying Enterprise Decision Intelligence with Generative AI in a Financial Services Company

Case	Action
Company Background	Wealth Management Financial Services Company is a mid-sized financial institution offering a range of services, including retail banking, investment management, and insurance. With a customer base of over 2 million, the company operates in a highly competitive market and strives to differentiate itself through innovative technology solutions.
Problem Statement	Wealth Management Financial Services Company faced several challenges: **1. Inefficient Decision-Making:** Decision-making processes were slow and often based on incomplete data, leading to missed opportunities and suboptimal outcomes. **2. Customer Experience:** The company struggled to provide personalized services, resulting in lower customer satisfaction and retention. **3. Operational Costs:** High operational costs due to manual processes and inefficiencies. **4. Fraud Detection:** Existing fraud detection systems were not effective, leading to significant financial losses.
Objectives	**1. Enhance Decision-Making:** Implement generative AI to provide real-time, data-driven insights. **2. Improve Customer Experience:** Use AI to personalize customer interactions and services. **3. Increase Operational Efficiency:** Automate routine tasks to reduce costs and improve efficiency. **4. Strengthen Fraud Detection:** Deploy advanced AI models to detect and prevent fraudulent activities.

Case	Action
Solution	Wealth Management Financial Services Company partnered with a leading AI technology provider to implement an enterprise decision intelligence platform powered by generative AI. The solution included: **1. AI-Powered Analytics Engine:** Integrated with the company's data infrastructure to analyze large datasets and generate actionable insights. **2. Customer Personalization Module:** Leveraged AI to tailor services and product recommendations based on individual customer behavior and preferences. **3. Automation Tools:** Automated routine tasks such as loan processing, customer support, and regulatory reporting. **4. Fraud Detection System:** Deployed AI models to identify suspicious transactions and potential fraud in real-time.
Implementation	**1. Phase 1: Assessment and Planning** • Conducted a thorough assessment of existing processes and identified areas for AI integration. • Developed a detailed implementation roadmap with clear milestones and timelines. **2. Phase 2: Pilot Project** • Initiated a pilot project focusing on fraud detection and customer personalization. • Collected feedback and refined the AI models based on pilot results. **3. Phase 3: Full-Scale Deployment** • Rolled out the AI-powered decision intelligence platform across all departments. • Provided extensive training for employees to ensure smooth adoption. **4. Phase 4: Continuous Improvem**Established a monitoring and optimization framework to continuously improve AI models and processes.

Case	Action
Results	**1. Enhanced Decision-Making** • Decision-making processes became significantly faster and more accurate, leading to better business outcomes. • The AI-powered analytics engine provided real-time insights, enabling proactive strategies and responses. **2. Improved Customer Experience** • Customer satisfaction scores increased by 20% due to personalized services and interactions. • The customer personalization module resulted in a 15% increase in cross-selling and upselling opportunities. **3. Operational Efficiency** • Automation of routine tasks led to a 30% reduction in operational costs. • Loan processing times were reduced by 50%, enhancing customer satisfaction and operational throughput. **4. Strengthened Fraud Detection** • The new fraud detection system reduced financial losses due to fraud by 40%. • The AI models identified suspicious activities with higher accuracy, leading to faster and more effective interventions.

Financial Impact

Year	Investment ($M)	Cost Savings ($M)	Revenue Growth ($M)	Net Benefit ($M)
1	8	3	2	-3
2	7	8	5	6
3	5	12	8	15
4	4	15	10	21
5	2	18	12	28

- **Total Investment**: $26 million
- **Total Cost Savings**: $56 million
- **Total Revenue Growth**: $37 million
- **Net Benefit**: $67 million

ROI Calculation

$$ROI = \frac{\text{Total Net Benefits} - \text{Total Costs}}{\text{Total Costs}} \times 100$$

- **Total Net Benefits**: $67 million
- **Total Costs**: $26 million

$$ROI = \frac{67 - 26}{26} \times 100 = 157.7\%$$

$\downarrow$

ROI Calculation

$$ROI = \frac{\text{Total Net Benefits} - \text{Total Costs}}{\text{Total Costs}} \times 100$$

- **Total Net Benefits**: $67 million
- **Total Costs**: $26 million

$$ROI = \frac{67 - 26}{26} \times 100 = 157.7\%$$

$\downarrow$

Conclusion

The implementation of enterprise decision intelligence with generative AI at Wealth Management Financial Services Company led to significant improvements in decision-making, customer experience, operational efficiency, and fraud detection. The financial impact was substantial, with a 157.7% ROI over five years. This case study demonstrates the transformative potential of generative AI in the financial services industry and serves as a model for other institutions aiming to leverage AI for strategic advantage.

Applying to CPG industry including a business case, ROI, SWOT Analysis and Case Study

Executive Summary

The CPG industry is characterized by intense competition, evolving consumer preferences, and complex supply chains. Generative AI offers a transformative approach to decision intelligence, enabling CPG companies to enhance their decision-making processes, improve consumer engagement, optimize supply chain management, and drive innovation. This business case outlines the strategic implementation of generative AI in the CPG industry, focusing on key benefits, potential use cases, implementation strategy, and expected outcomes.

Problem Statement

CPG companies face several challenges, including the need for real-time decision-making, demand forecasting, personalized marketing, and efficient supply chain management. Traditional analytical methods often fall short in addressing these complexities. There is a pressing need for advanced solutions that can handle large volumes of data, provide actionable insights, and support strategic decision-making.

Objectives

1. Enhance Decision-Making: Utilize generative AI to provide real-time, data-driven insights for better decision-making.

2. Improve Consumer Engagement: Personalize marketing and product offerings to meet individual consumer preferences.

3. Optimize Supply Chain: Streamline supply chain operations to reduce costs and improve efficiency.

4. Drive Innovation: Accelerate product development and innovation through predictive analytics and trend analysis.

Key Benefits

1. Data-Driven Insights: Generative AI can process vast amounts of data to uncover patterns and insights that humans might miss, leading to more informed decisions.

2. Personalization: Tailor marketing campaigns and product offerings to individual consumer preferences, enhancing engagement and loyalty.

3. Supply Chain Optimization: Improve demand forecasting, inventory management, and logistics to reduce costs and improve efficiency.

4. Cost Reduction: Automate routine tasks to reduce operational costs and increase efficiency.

5. Innovation Acceleration: Identify emerging trends and consumer needs to drive product innovation and development.

Use Cases
1. Demand Forecasting: Use AI to predict consumer demand with high accuracy, optimizing inventory levels and reducing stockouts or overstock situations.

2. Personalized Marketing: Leverage AI to create targeted marketing campaigns based on consumer behavior and preferences.

3. Supply Chain Management: Implement AI-driven solutions for efficient supply chain planning, logistics, and risk management.

4. Product Development: Utilize AI to analyze market trends and consumer feedback for new product development and innovation.

5. Customer Service: Implement AI-powered chatbots and virtual assistants to enhance customer support and engagement.

Implementation Strategy

1. Assessment and Planning
- Conduct a comprehensive assessment of current decision-making processes and identify areas for AI integration.
- Develop a strategic roadmap outlining the implementation phases, resource requirements, and key milestones.

2. Pilot Projects
- Initiate pilot projects in selected areas such as demand forecasting or personalized marketing to demonstrate the value of generative AI.
- Gather feedback and refine models before broader deployment.

3. Technology Integration
- Integrate generative AI solutions with existing IT infrastructure, ensuring compatibility and data security.
- Train employees on new tools and workflows to facilitate smooth adoption.

4. Scalability and Optimization
- Scale successful pilot projects across the organization.
- Continuously monitor and optimize AI models to improve performance and adapt to changing business needs.

Risks and Mitigation
1. Data Privacy and Security: Implement robust data protection measures and comply with data privacy regulations.

2. Integration Challenges: Ensure seamless integration with existing systems through comprehensive planning and testing.

3. Resistance to Change: Foster a culture of innovation and provide training to ease the transition for employees.

4. Model Accuracy: Continuously validate and update AI models to maintain accuracy and relevance.

Expected Outcomes
1. Enhanced Decision-Making: Faster, more accurate decisions leading to improved business performance.

2. Improved Consumer Engagement: Higher consumer satisfaction and retention through personalized marketing.

3. Supply Chain Efficiency: Significant cost savings and efficiency gains through optimized supply chain management.

Financial Projections

Year	Investment ($M)	Cost Savings ($M)	Revenue Growth ($M)
1	12	6	3
2	10	12	7
3	8	18	12
4	6	24	18
5	4	30	24

ROI Calculation

$$\text{ROI} = \frac{\text{Total Net Benefits} - \text{Total Costs}}{\text{Total Costs}} \times 100$$

- **Total Net Benefits** over 5 years: $132 million

- **Total Costs** (Investment + Operational Costs): $40 million

$$\text{ROI} = \frac{132-40}{40} \times 100 = 230\%$$

Business Case: Enterprise Decision Intelligence with Generative AI for the CPG Industry

Business Case for Implementing Enterprise Strategy Intelligence

1. Executive Summary:

Enterprise Decision Intelligence (EDI) with Generative AI has the potential to revolutionize decision-making processes in the Consumer Packaged Goods (CPG) industry. By leveraging advanced analytics, predictive modeling, and Generative AI algorithms, organizations can drive product innovation, optimize operations, and improve customer experiences. This business case outlines the benefits and value proposition of implementing EDI with Generative AI in the CPG industry.

2. Problem Statement:

CPG companies face challenges in optimizing their product portfolios, forecasting demand accurately, and meeting evolving consumer preferences. Traditional decision-making processes often lack the ability to analyze large datasets, extract valuable insights, and adapt quickly to changing market dynamics.

3. Solution Overview:

EDI with Generative AI enables organizations to harness the power of data and advanced analytics to improve decision-making in the CPG industry. By integrating diverse data sources, applying AI algorithms, and generating insights, organizations can make informed, data-driven decisions that drive product innovation, optimize supply chain operations, and enhance customer experiences.

4. Benefits and Value Proposition:

a. Product Innovation: EDI with Generative AI leverages consumer data, market trends, and competitor insights to generate new product ideas, identify market gaps, and enhance product development processes. This leads to the creation of innovative products that align with consumer preferences and drive revenue growth.

b. Demand Forecasting and Inventory Optimization: EDI with Generative AI enables accurate demand forecasting by analyzing historical sales data, market trends, and external factors. This optimizes inventory management, reduces stockouts, and minimizes inventory holding costs.

c. Personalized Marketing and Customer Experience: EDI with Generative AI leverages consumer data, including demographics, preferences, and purchase history, to deliver personalized marketing campaigns and enhance customer experiences. This improves customer engagement, loyalty, and sales conversion rates.

d. Operational Efficiency: EDI with Generative AI optimizes supply chain operations by analyzing data on production, transportation, and distribution. This streamlines processes, reduces costs, and enhances overall operational efficiency.

5. ROI and Financial Impact:

a. Cost Savings: EDI with Generative AI optimizes inventory levels, reduces stockouts, and minimizes holding costs. This leads to cost savings through improved supply chain efficiency and reduced waste.

b. Revenue Growth: Improved product innovation, accurate demand forecasting, and personalized marketing drive revenue growth by capturing market share, increasing customer loyalty, and driving repeat purchases.

6. Implementation Strategy:

a. Data Integration: Establish a robust data integration framework to aggregate diverse CPG data, including sales data, customer data, market trends, and competitor insights.

b. AI Model Development: Develop Generative AI models tailored to specific CPG objectives, such as new product development, demand forecasting, and customer segmentation.

c. Change Management: Implement a comprehensive change management plan to foster a data-driven culture, train employees on EDI with Generative AI, and address potential resistance to change.

d. Collaboration: Foster cross-functional collaboration between product development, marketing, supply chain, and data science teams to effectively implement and leverage EDI with Generative AI.

7. Risks and Mitigation Strategies:

a. Data Privacy and Ethics: Implement robust data privacy protocols, ensure compliance with regulations, and address ethical considerations to protect consumer information and mitigate potential risks.

b. Data Availability and Quality: Establish partnerships with data providers, invest in data quality assurance processes, and leverage advanced data cleansing techniques to ensure the availability of accurate and reliable data.

8. Conclusion:

Implementing Enterprise Decision Intelligence with Generative AI in the CPG industry offers significant benefits and potential ROI. By leveraging advanced analytics, optimizing product portfolios, enhancing demand forecasting, and personalizing marketing efforts, organizations can drive growth, increase operational efficiency, and meet evolving consumer demands. Through effective implementation, organizations can establish themselves as leaders in the CPG market, capitalize on emerging opportunities, and create long-term value.

ROI (Return on Investment) Analysis: Enterprise Decision Intelligence with Generative AI for the CPG Industry

Introduction:

Implementing Enterprise Decision Intelligence (EDI) with Generative AI in the Consumer Packaged Goods (CPG) industry has the potential to generate significant returns on investment. By leveraging advanced analytics, predictive modeling, and Generative AI algorithms, CPG companies can drive product innovation, optimize operations, and improve financial performance. This ROI analysis explores the potential financial benefits of implementing EDI with Generative AI in the CPG industry.

Case	Action
Cost Savings:	**a. Supply Chain Optimization:** EDI with Generative AI helps optimize supply chain operations by analyzing data on inventory levels, transportation, and demand forecasts. This reduces carrying costs, minimizes stockouts, and eliminates excess stock, resulting in cost savings. **b. Inventory Management:** Improved demand forecasting and inventory optimization through Generative AI algorithms minimize inventory holding costs and reduce the risk of overstocking or stockouts. This leads to lower carrying costs, improved cash flow, and reduced waste. **c. Operational Efficiency:** Streamlining processes, automating tasks, and optimizing resource allocation with EDI and Generative AI enhance operational efficiency, reducing labor costs and improving productivity.
Revenue Growth:	**a. Product Innovation:** EDI with Generative AI leverages consumer data, market trends, and competitor insights to generate new product ideas and identify market gaps. This drives product innovation, which can lead to increased sales and market share. **b. Personalized Marketing:** EDI with Generative AI enables personalized marketing campaigns by leveraging consumer data, preferences, and purchase history. This improves customer engagement, loyalty, and sales conversion rates, resulting in increased revenue.

Case	Action
	c. Pricing Optimization: Generative AI algorithms analyze pricing dynamics, market conditions, and consumer behavior to optimize pricing strategies. This maximizes revenue, enhances competitiveness, and improves profitability.
Competitive Advantage:	**a. Enhanced Customer Experience:** Implementing EDI with Generative AI enables CPG companies to provide personalized, engaging customer experiences. This enhances customer satisfaction, loyalty, and brand advocacy, giving the organization a competitive advantage. **b. Market Agility:** EDI with Generative AI enables real-time data analysis, allowing CPG companies to quickly respond to changing market trends and consumer preferences. This agility helps seize new opportunities and respond to market disruptions faster, staying ahead of competitors.
Cost of Implementation and Maintenance:	**a. Initial Investment:** Implementing EDI with Generative AI requires an initial investment in infrastructure, technology, and talent acquisition. Costs may include software licenses, hardware, data integration, and training programs. **b. Ongoing Maintenance:** Ongoing costs include system updates, data management, algorithm fine-tuning, and personnel training to ensure optimal performance.
ROI Calculation:	To calculate ROI, compare the financial benefits (cost savings and revenue growth) against the investment costs: ROI (%) = ((Financial Benefits - Investment Costs) / Investment Costs) x 100 Note: The ROI calculation should consider the specific costs and benefits relevant to the organization's context and should be based on realistic projections and assumptions.

Case	Action
Conclusion	Enterprise Decision Intelligence with Generative AI offers substantial potential for ROI in the CPG industry. By optimizing supply chain operations, improving inventory management, driving revenue growth through roduct innovation and personalized marketing, and gaining a competitive advantage, CPG companies can achieve significant financial benefits. Although there are initial investment and ongoing maintenance costs, the potential returns on investment make the implementation of EDI with Generative AI a compelling proposition for CPG companies seeking to enhance operational efficiency, drive growth, and improve their financial performance.

SWOT Analysis: Decision Intelligence with Generative AI for the CPG Industry

Decision Intelligence with Generative AI presents significant strengths and opportunities for CPG companies, including enhanced decision-making, product innovation, demand forecasting, and personalized marketing. However, challenges related to data availability, ethical considerations, data security, and competition must be effectively managed.
By leveraging the strengths, addressing weaknesses, capitalizing on opportunities, and mitigating threats, CPG companies can harness the full potential of Decision Intelligence with Generative AI to drive growth, optimize supply chain operations, and meet evolving consumer demands.

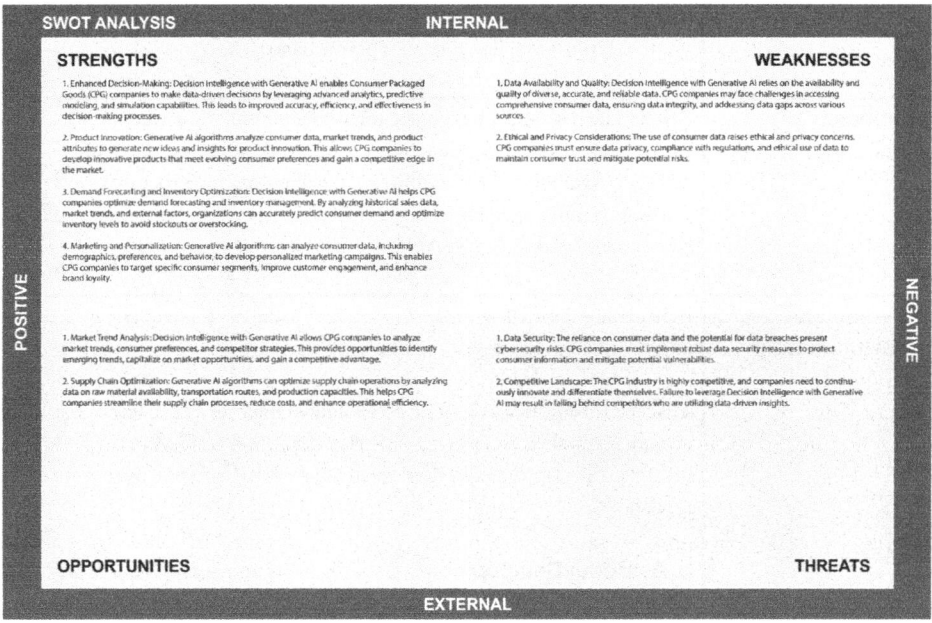

SWOT ANALYSIS — **INTERNAL**

STRENGTHS

1. Enhanced Decision-Making: Decision Intelligence with Generative AI enables Consumer Packaged Goods (CPG) companies to make data-driven decisions by leveraging advanced analytics, predictive modeling, and simulation capabilities. This leads to improved accuracy, efficiency, and effectiveness in decision-making processes.

2. Product Innovation: Generative AI algorithms analyze consumer data, market trends, and product attributes to generate new ideas and insights for product innovation. This allows CPG companies to develop innovative products that meet evolving consumer preferences and gain a competitive edge in the market.

3. Demand Forecasting and Inventory Optimization: Decision Intelligence with Generative AI helps CPG companies optimize demand forecasting and inventory management. By analyzing historical sales data, market trends, and external factors, organizations can accurately predict consumer demand and optimize inventory levels to avoid stockouts or overstocking.

4. Marketing and Personalization: Generative AI algorithms can analyze consumer data, including demographics, preferences, and behavior, to develop personalized marketing campaigns. This enables CPG companies to target specific consumer segments, improve customer engagement, and enhance brand loyalty.

WEAKNESSES

1. Data Availability and Quality: Decision Intelligence with Generative AI relies on the availability and quality of diverse, accurate, and reliable data. CPG companies may face challenges in accessing comprehensive consumer data, ensuring data integrity, and addressing data gaps across various sources.

2. Ethical and Privacy Considerations: The use of consumer data raises ethical and privacy concerns. CPG companies must ensure data privacy, compliance with regulations, and ethical use of data to maintain consumer trust and mitigate potential risks.

POSITIVE / **NEGATIVE**

1. Market Trend Analysis: Decision Intelligence with Generative AI allows CPG companies to analyze market trends, consumer preferences, and competitor strategies. This provides opportunities to identify emerging trends, capitalize on market opportunities, and gain a competitive advantage.

2. Supply Chain Optimization: Generative AI algorithms can optimize supply chain operations by analyzing data on raw material availability, transportation routes, and production capacities. This helps CPG companies streamline their supply chain processes, reduce costs, and enhance operational efficiency.

1. Data Security: The reliance on consumer data and the potential for data breaches present cybersecurity risks. CPG companies must implement robust data security measures to protect consumer information and mitigate potential vulnerabilities.

2. Competitive Landscape: The CPG industry is highly competitive, and companies need to continuously innovate and differentiate themselves. Failure to leverage Decision Intelligence with Generative AI may result in falling behind competitors who are utilizing data-driven insights.

OPPORTUNITIES — **THREATS**

EXTERNAL

Case Study: Enterprise Decision Intelligence with Generative AI for the CPG Industry

Company: Beverage Based Consumer Goods (BBCG)
Industry: Consumer Packaged Goods (CPG)

Case	Action
Introduction:	Beverage Based Consumer Goods (BBCG) is a leading CPG company known for its diverse product portfolio in the food and beverage industry. To stay competitive and meet evolving consumer demands, BBCG implemented Enterprise Decision Intelligence (EDI) with Generative AI. This case study highlights the successful implementation of EDI with Generative AI and its impact on BBCG's operations, product innovation, and financial performance.
Business Challenge:	BBCG faced several challenges, including optimizing product assortments, forecasting demand accurately, and staying ahead of emerging market trends. Traditional decision-making processes were time-consuming and lacked the ability to analyze vast amounts of data and extract actionable insights.
Solution and Implementation:	BBCG adopted EDI with Generative AI to address these challenges. The implementation involved the following steps: **a. Data Integration:** BBCG consolidated internal and external data sources, including sales data, customer data, market trends, competitor insights, and social media data, to create a comprehensive data repository. **b. AI Model Development:** Data scientists and analysts developed Generative AI models tailored to specific objectives, such as demand forecasting, market trend analysis, and product innovation. These models utilized advanced analytics and machine learning algorithms to generate insights and recommendations. **c. Infrastructure and Training:** BBCG invested in the necessary technology infrastructure, software, and hardware to support the EDI system. Employees across various departments, including marketing, supply chain, and product development, received training on data-driven decision-making and utilizing the EDI platform.

Case	Action
Implementation Results:	**a. Demand Forecasting:** The Generative AI models improved demand forecasting accuracy by analyzing historical sales data, market trends, and external factors. This enabled BBCG to optimize inventory levels, reduce stockouts, and align production with actual consumer demand. **b. Product Innovation:** By leveraging Generative AI algorithms, BBCG gained insights into emerging market trends, consumer preferences, and competitor offerings. This facilitated the development of innovative products, addressing changing consumer demands and driving market differentiation. **c. Pricing Optimization:** Generative AI models analyzed pricing dynamics, competitor pricing strategies, and market conditions to optimize pricing decisions. This resulted in improved pricing competitiveness, increased sales, and enhanced profitability. **d. Customer Segmentation and Personalization:** The EDI system enabled BBCG to segment its customer base and deliver personalized marketing campaigns. By understanding consumer preferences and behaviors, BBCG tailored its marketing efforts, leading to higher customer engagement, improved customer loyalty, and increased sales conversion rates.
Financial Impact:	**a. Revenue Growth:** The implementation of EDI with Generative AI resulted in increased revenue for BBCG. Improved demand forecasting, product innovation, and personalized marketing campaigns drove sales growth and market share expansion. **b. Cost Savings:** Optimized inventory management, streamlined supply chain operations, and efficient resource allocation led to cost savings. Reduced stockouts, improved production planning, and lower inventory carrying costs contributed to improved profitability.

Case	Action
Competitive Advantage:	BBCG gained a competitive advantage through its implementation of EDI with Generative AI. The ability to leverage data-driven insights for decision-making, respond quickly to market trends, and deliver personalized customer experiences positioned BBCG as an industry leader.
Conclusion	The implementation of Enterprise Decision Intelligence with Generative AI enabled Beverage Based Consumer Goods to overcome challenges, drive growth, and achieve competitive advantage in the CPG industry. By leveraging advanced analytics, demand forecasting, product innovation, and personalized marketing, BBCG experienced revenue growth, cost savings, and improved operational efficiency. The success of the EDI implementation showcases the potential of Generative AI in revolutionizing decision-making processes and driving business performance in the CPG sector.

Applying to Retail Industry including a Business Case, ROI, SWOT Analysis and Case Study

Executive Summary
The retail industry faces numerous challenges such as dynamic consumer preferences, competitive pressures, and complex supply chain logistics. Generative AI offers advanced decision intelligence capabilities that can significantly enhance decision-making processes, personalize consumer experiences, optimize supply chain management, and drive innovation. This business case explores the strategic implementation of generative AI in the retail industry, detailing the key benefits, potential use cases, implementation strategy, and expected outcomes.

Problem Statement
Retailers often struggle with the following challenges:
1. **Dynamic Consumer Preferences:** Rapid changes in consumer behavior and preferences.
2. **Operational Inefficiencies:** Inefficiencies in supply chain and inventory management.
3. **Competitive Pressures:** Intense competition requiring innovative and personalized customer engagement strategies.
4. **Overload:** Handling and deriving actionable insights from vast amounts of data.

Objectives
1. **Enhance Decision-Making:** Utilize generative AI to provide real-time, data-driven insights for better decision-making.
2. **Improve Consumer Engagement:** Personalize marketing and shopping experiences to boost consumer satisfaction and loyalty.
3. **Optimize Supply Chain:** Streamline supply chain operations to reduce costs and improve efficiency.
4. **Drive Innovation:** Leverage AI for product development and market trend analysis.

Key Benefits
1. **Data-Driven Insights:** Generative AI can process large datasets to uncover patterns and trends, enhancing decision-making.
2. **Personalization:** Tailor marketing campaigns and shopping experiences to individual consumer preferences.
3. **Supply Chain Optimization:** Improve demand forecasting, inventory management, and logistics to reduce costs and improve efficiency.
4. **Cost Reduction:** Automate routine tasks to reduce operational costs and increase efficiency.
5. **Innovation Acceleration:** Identify emerging trends and consumer needs to drive product innovation and development.

Use Cases
1. **Personalized Marketing:** Use AI to create targeted marketing campaigns based on consumer behavior and preferences.
2. **Demand Forecasting:** Leverage AI to predict consumer demand accurately, optimizing inventory levels and reducing stockouts.
3. **Supply Chain Management:** Implement AI-driven solutions for efficient supply chain

planning, logistics, and risk management.

4. **Customer Service:** Deploy AI-powered chatbots and virtual assistants to enhance customer support and engagement.

5. **Product Recommendations:** Use AI to analyze purchase history and preferences to provide personalized product recommendations.

1. Assessment and Planning

- Conduct a comprehensive assessment of current decision-making processes and identify areas for AI integration.
- Develop a strategic roadmap outlining the implementation phases, resource requirements, and key milestones.

2. Pilot Projects

- Initiate pilot projects in selected areas such as personalized marketing or demand forecasting to demonstrate the value of generative AI.
- Gather feedback and refine models before broader deployment.

3. Technology Integration

- Integrate generative AI solutions with existing IT infrastructure, ensuring compatibility and data security.
- Train employees on new tools and workflows to facilitate smooth adoption.

4. Scalability and Optimization

- Scale successful pilot projects across the organization.
- Continuously monitor and optimize AI models to improve performance and adapt to changing business needs.

Risks and Mitigation

1. **Data Privacy and Security:** Implement robust data protection measures and comply with data privacy regulations.
2. **Integration Challenges:** Ensure seamless integration with existing systems through comprehensive planning and testing.
3. **Resistance to Change:** Foster a culture of innovation and provide training to ease the transition for employees.
4. **Model Accuracy:** Continuously validate and update AI models to maintain accuracy and relevance.

Expected Outcomes

1. Enhanced Decision-Making: Faster, more accurate decisions leading to improved business performance.
2. Improved Consumer Engagement: Higher consumer satisfaction and retention through personalized marketing and shopping experiences.
3. Supply Chain Efficiency: Significant cost savings and efficiency gains through optimized supply chain management.
4. Innovation and Growth: Accelerated product development and market growth through data-driven insights.

Financial Impact

Year	Investment ($M)	Cost Savings ($M)	Revenue Growth ($M)	Net Benefit ($M)
1	15	8	5	-2
2	12	16	10	14
3	10	24	15	29
4	8	32	20	44
5	6	40	25	59

- **Total Investment**: $51 million

- **Total Cost Savings**: $120 million

- **Total Revenue Growth**: $75 million

- **Net Benefit**: $144 million

ROI Calculation

$$ROI = \frac{\text{Total Net Benefits} - \text{Total Costs}}{\text{Total Costs}} \times 100$$

- **Total Net Benefits**: $144 million

- **Total Costs**: $51 million

$$ROI = \frac{144 - 51}{51} \times 100 = 182.4\%$$

ROI (Return on Investment) Analysis: Enterprise Decision Intelligence with Generative AI for the Retail Industry

Introduction:

Implementing Enterprise Decision Intelligence (EDI) with Generative AI in the retail industry has the potential to deliver significant returns on investment. By leveraging advanced analytics, predictive modeling, and Generative AI algorithms, retail organizations can drive customer-centric strategies, optimize operations, and improve financial performance. This ROI analysis explores the potential financial benefits of implementing EDI with Generative AI in the retail industry.

Case	Action
Cost Savings:	**a. Supply Chain Optimization:** EDI with Generative AI helps optimize supply chain operations by analyzing data on inventory levels, transportation, and demand forecasts. This reduces carrying costs, minimizes stockouts, and eliminates excess stock, resulting in cost savings. **b. Inventory Management:** Improved demand forecasting and inventory optimization through Generative AI algorithms minimize inventory holding costs and reduce the risk of overstocking or stockouts. This leads to lower carrying costs, improved cash flow, and reduced waste. **c. Operational Efficiency:** Streamlining processes, automating tasks, and optimizing resource allocation with EDI and Generative AI enhance operational efficiency, reducing labor costs and improving productivity.
Revenue Growth:	**a. Personalized Marketing:** EDI with Generative AI enables personalized marketing campaigns by leveraging consumer data, preferences, and purchase history. This improves customer engagement, loyalty, and sales conversion rates, resulting in increased revenue. **b. Pricing Optimization:** Generative AI algorithms analyze pricing dynamics, market conditions, and consumer behavior to optimize pricing strategies. This maximizes revenue, enhances competitiveness, and improves profitability.

Case	Action
	c. Product Assortment and Innovation: Generative AI algorithms generate insights on market trends, consumer preferences, and competitor offerings. This enables retail organizations to optimize product assortments, introduce innovative products, and capture new market segments, driving revenue growth.
Competitive Advantage:	**a. Enhanced Customer Experience:** Implementing EDI with Generative AI enables retail organizations to provide personalized, seamless customer experiences. This enhances customer satisfaction, loyalty, and retention, giving the organization a competitive advantage. **b. Agility and Adaptability:** EDI with Generative AI enables real-time data analysis, allowing retail organizations to quickly adapt to changing market trends and consumer preferences. This agility helps seize new opportunities and respond to market disruptions faster, staying ahead of competitors.
Cost of Implementation and Maintenance:	**a. Initial Investment:** Implementing EDI with Generative AI requires an initial investment in infrastructure, technology, and talent acquisition. Costs may include software licenses, hardware, data integration, and training programs. **b. Ongoing Maintenance:** Ongoing costs include system updates, data management, algorithm fine-tuning, and personnel training to ensure optimal performance.
ROI Calculation:	To calculate ROI, compare the financial benefits (cost savings and revenue growth) against the investment costs: ROI (%) = ((Financial Benefits - Investment Costs) / Investment Costs) x 100 Note: The ROI calculation should consider the specific costs and benefits relevant to the organization's context and should be based on realistic projections and assumptions.

Case	Action
Conclusion	Enterprise Decision Intelligence with Generative AI offers substantial potential for ROI in the retail industry. By optimizing supply chain operations, improving inventory management, driving revenue growth through personalized marketing and pricing optimization, and gaining a competitive advantage, retail organizations can achieve significant financial benefits. Although there are initial investment and ongoing maintenance costs, the potential returns on investment make the implementation of EDI with Generative AI a compelling proposition for retailers seeking to enhance operational efficiency, drive growth, and improve their financial performance.

SWOT Analysis: Decision Intelligence with Generative AI for the Retail Industry

Decision Intelligence with Generative AI presents significant strengths and opportunities for the retail industry, including enhanced decision-making, demand forecasting, personalized marketing, and product assortment. However, challenges related to data availability, ethical considerations, data security, and competition must be effectively managed. By leveraging the strengths, addressing weaknesses, capitalizing on opportunities, and mitigating threats, retail organizations can harness the full potential of Decision Intelligence with Generative AI to drive growth, optimize customer experiences, and stay ahead in a rapidly evolving retail landscape.

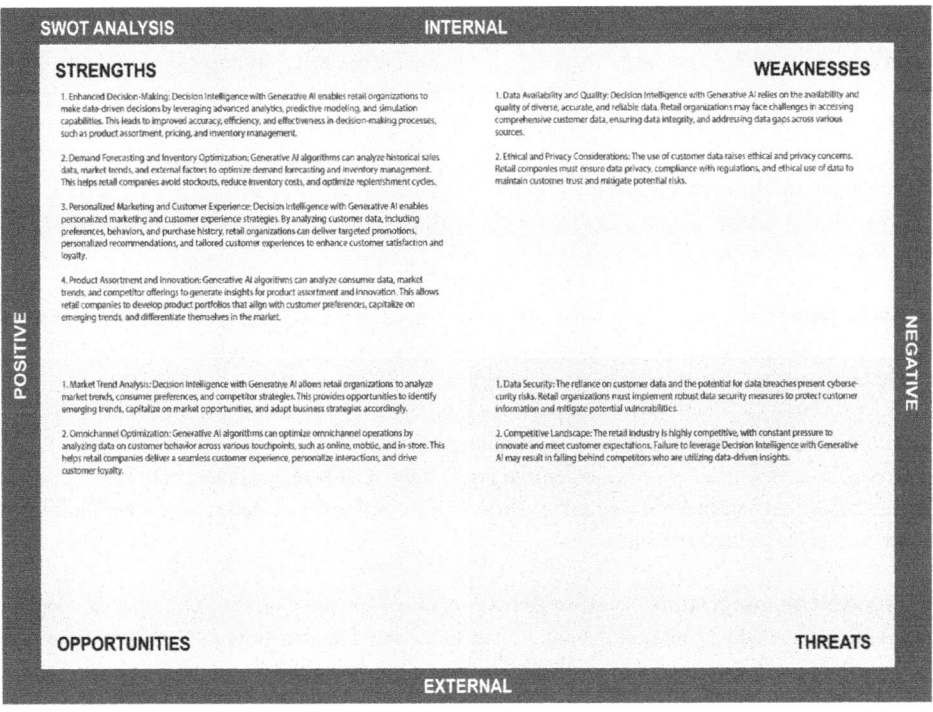

SWOT ANALYSIS — **INTERNAL**

STRENGTHS

1. Enhanced Decision-Making: Decision Intelligence with Generative AI enables retail organizations to make data-driven decisions by leveraging advanced analytics, predictive modeling, and simulation capabilities. This leads to improved accuracy, efficiency, and effectiveness in decision-making processes, such as product assortment, pricing, and inventory management.

2. Demand Forecasting and Inventory Optimization: Generative AI algorithms can analyze historical sales data, market trends, and external factors to optimize demand forecasting and inventory management. This helps retail companies avoid stockouts, reduce inventory costs, and optimize replenishment cycles.

3. Personalized Marketing and Customer Experience: Decision Intelligence with Generative AI enables personalized marketing and customer experience strategies. By analyzing customer data, including preferences, behaviors, and purchase history, retail organizations can deliver targeted promotions, personalized recommendations, and tailored customer experiences to enhance customer satisfaction and loyalty.

4. Product Assortment and Innovation: Generative AI algorithms can analyze consumer data, market trends, and competitor offerings to generate insights for product assortment and innovation. This allows retail companies to develop product portfolios that align with customer preferences, capitalize on emerging trends, and differentiate themselves in the market.

WEAKNESSES

1. Data Availability and Quality: Decision Intelligence with Generative AI relies on the availability and quality of diverse, accurate, and reliable data. Retail organizations may face challenges in accessing comprehensive customer data, ensuring data integrity, and addressing data gaps across various sources.

2. Ethical and Privacy Considerations: The use of customer data raises ethical and privacy concerns. Retail companies must ensure data privacy, compliance with regulations, and ethical use of data to maintain customer trust and mitigate potential risks.

POSITIVE / **NEGATIVE**

1. Market Trend Analysis: Decision Intelligence with Generative AI allows retail organizations to analyze market trends, consumer preferences, and competitor strategies. This provides opportunities to identify emerging trends, capitalize on market opportunities, and adapt business strategies accordingly.

2. Omnichannel Optimization: Generative AI algorithms can optimize omnichannel operations by analyzing data on customer behavior across various touchpoints, such as online, mobile, and in-store. This helps retail companies deliver a seamless customer experience, personalize interactions, and drive customer loyalty.

1. Data Security: The reliance on customer data and the potential for data breaches present cybersecurity risks. Retail organizations must implement robust data security measures to protect customer information and mitigate potential vulnerabilities.

2. Competitive Landscape: The retail industry is highly competitive, with constant pressure to innovate and meet customer expectations. Failure to leverage Decision Intelligence with Generative AI may result in falling behind competitors who are utilizing data-driven insights.

OPPORTUNITIES — **THREATS**

EXTERNAL

Case Study: Enterprise Decision Intelligence with Generative AI for the Retail Industry

Company: Footwear Retail Corporation
Industry: Retail

1. Introduction:
Footwear Retail Corporation is a global retail giant operating a chain of stores across various regions. To maintain its competitive edge and navigate the rapidly evolving retail landscape, Footwear Retail Corporation adopted Enterprise Decision Intelligence (EDI) with Generative AI. This case study highlights the successful implementation of EDI with Generative AI and its impact on Footwear Retail Corporation's operations, customer experience, and financial performance.

2. Business Challenge:
Footwear Retail Corporation faced challenges in optimizing product assortments, pricing strategies, and customer experiences. Traditional decision-making processes relied heavily on manual analysis and lacked the agility to respond to changing market dynamics and customer preferences.

3. Solution and Implementation:
Footwear Retail Corporation implemented EDI with Generative AI to address these challenges. The implementation involved the following steps:

a. Data Integration: Footwear Retail Corporation consolidated and integrated data from various sources, including point-of-sale data, customer data, inventory data, market trends, and social media sentiment analysis, to create a comprehensive data infrastructure.

b. AI Model Development: Data scientists and analysts developed Generative AI models tailored to specific objectives, such as demand forecasting, pricing optimization, customer segmentation, and personalized marketing. These models utilized machine learning algorithms to generate insights and recommendations.

c. Technology Integration: Footwear Retail Corporation invested in the necessary technology infrastructure, including cloud computing, AI platforms, and data analytics tools, to support the EDI system. The system was integrated with existing IT systems to ensure seamless data flow and decision-making integration.

4. Implementation Results:
a. Demand Forecasting and Inventory Optimization: The Generative AI models improved demand forecasting accuracy by analyzing historical sales data, market trends, and external factors. This enabled Footwear Retail Corporation to optimize inventory levels, reduce stockouts, and minimize excess inventory, leading to improved operational efficiency and cost savings.

b. Pricing Optimization: The EDI system, powered by Generative AI, analyzed pricing dynamics, competitor pricing strategies, and market conditions to optimize pricing decisions. Footwear Retail Corporation was able to set competitive prices, maximize revenue, and enhance profitability.

c. Customer Segmentation and Personalized Marketing: Generative AI models enabled Footwear Retail Corporation to segment its customer base and deliver personalized marketing campaigns. By analyzing customer data, preferences, and purchase history, Footwear Retail Corporation tailored its marketing efforts, leading to increased customer engagement, loyalty, and sales conversion rates.

d. Real-time Analytics and Insights: The EDI system provided real-time analytics and insights to store managers, enabling them to make data-driven decisions on product placements, promotions, and operational adjustments. This enhanced agility, improved decision-making, and optimized store performance.

5. Financial Impact:
a. Revenue Growth: The implementation of EDI with Generative AI resulted in increased revenue for Footwear Retail Corporation. Improved demand forecasting, optimized pricing strategies, and personalized marketing campaigns drove sales growth, customer acquisition, and market share expansion.

b. Cost Savings: By optimizing inventory levels, reducing stockouts, and streamlining pricing strategies, Footwear Retail Corporation achieved significant cost savings. Operational efficiency improvements and better resource allocation led to reduced carrying costs and improved profitability.

6. Competitive Advantage:
Footwear Retail Corporation gained a competitive advantage through its implementation of EDI with Generative AI. The ability to leverage data-driven insights, optimize pricing, personalize marketing, and make informed decisions enabled Footwear Retail Corporation to stay ahead of competitors, meet customer expectations, and adapt to market changes.

7. Conclusion:
The implementation of Enterprise Decision Intelligence with Generative AI empowered Footwear Retail Corporation to overcome challenges, drive growth, and gain a competitive edge in the retail industry. By leveraging advanced analytics, demand forecasting, pricing optimization, and personalized marketing, Footwear Retail Corporation experienced revenue growth, cost savings, and improved operational efficiency. The success of the EDI implementation demonstrates the potential of Generative AI in revolutionizing decision-making processes and enhancing business performance in the retail sector.

Other industries
Of course, there are many other industries to apply this to. I just chose a top few to illustrate the power of Enterprise Decision Intelligence with Generative AI.

End of Section 2:
Recapping Section 2: Conclusion

The benefits of applying generative AI for enterprise decision intelligence are vast and transformative. From enhancing decision-making and driving innovation to optimizing supply chain management and personalizing customer experiences, generative AI offers powerful solutions that help industries stay competitive in a rapidly changing market. As businesses continue to embrace AI technologies, those that effectively integrate generative AI into their operations will be well-positioned to lead in their respective markets, driving growth and innovation in the years to come.

"Rather than learning how to solve that, shouldn't we be learning how to operate software that can solve that problem?"

Figure 16

Chapter 15: Starting Your Journey with Generative AI for Enterprise Decision Intelligence

Embarking on Your Enterprise Strategy Intelligence Journey with Generative AI

Introduction
In the modern business landscape, staying competitive and achieving success requires strategic decision-making based on accurate and timely information. Enterprise Strategy Intelligence (ESI) serves as a framework that enables organizations to gather, analyze, and leverage data-driven insights to make informed business choices. With the advancements in artificial intelligence (AI), particularly Generative AI, organizations now have a powerful tool at their disposal to enhance their ESI capabilities. This article explores how to embark on your journey with Enterprise Strategy Intelligence using Generative AI.

Understanding Enterprise Strategy Intelligence
Enterprise Strategy Intelligence refers to the process of systematically collecting, analyzing, and interpreting data to generate actionable insights for guiding business strategies and decision-

making. It encompasses a range of activities, including competitive intelligence, market analysis, trend forecasting, risk assessment, and opportunity identification. By harnessing ESI, organizations can gain a comprehensive understanding of their internal and external environments, enabling them to make informed strategic choices and adapt to dynamic market conditions.

Enter Generative AI

Generative AI, a branch of artificial intelligence, has rapidly evolved in recent years and is revolutionizing various industries. Unlike traditional AI, which focuses on pattern recognition and prediction, Generative AI models learn from vast amounts of data to generate new content, such as images, text, and even entire scenarios. This ability to generate new and original content makes Generative AI an invaluable tool for enterprise strategy intelligence.

Starting Your Journey

1. **Define your strategic goals:** Begin by clearly outlining your organization's strategic goals and the specific challenges you aim to address with ESI. Identify the key areas where Generative AI can provide valuable insights, such as market trends, customer behavior, product development, or risk assessment.
2. **Identify relevant data sources:** Determine the data sources required to fuel your Generative AI models. These sources may include internal data (e.g., sales figures, customer feedback) and external data (e.g., market research, social media analytics). Ensure the data is comprehensive, up-to-date, and relevant to your strategic goals.
3. **Build a Generative AI model:** Collaborate with AI experts or data scientists to develop a Generative AI model tailored to your ESI needs. This involves training the model on your data to generate insights and predictions. Consider utilizing pre-trained models or working with AI platforms that offer user-friendly tools for model development.
4. **Generate insights and validate:** Utilize your Generative AI model to generate insights and predictions based on the data you've provided. Assess the generated insights against existing knowledge, industry expertise, and historical data to validate their accuracy and relevance. Continuously refine and improve your model to enhance its predictive capabilities.
5. **Integrate ESI into decision-making processes:** Once you have a validated Generative AI model generating accurate insights, integrate these insights into your decision-making processes. Collaborate with key stakeholders, such as executives and department heads, to ensure that ESI becomes an integral part of strategic discussions and planning.
6. **Monitor, evaluate, and adapt:** Continuously monitor and evaluate the impact of ESI on your strategic decision-making processes. Measure the effectiveness of the insights generated by Generative AI against desired outcomes. Use feedback loops to refine your models and ensure they adapt to changing business landscapes.

Benefits and Considerations

Embracing Generative AI in your ESI journey offers numerous benefits, including:

1. **Enhanced strategic decision-making:** Generative AI provides organizations with deeper insights, enabling them to make more informed and data-driven strategic decisions.
2. **Agility and adaptability:** By leveraging Generative AI, organizations can quickly adapt to changing market conditions, identify emerging trends, and capitalize on new opportunities.

3. **Competitive advantage:** With access to cutting-edge technology, organizations can gain a competitive edge by staying ahead of industry trends, understanding customer needs, and making proactive strategic moves.

However, it is crucial to consider the ethical implications and potential challenges associated with Generative AI, such as data privacy, bias mitigation, and model explainability. Organizations must establish robust governance frameworks and ensure transparency when utilizing AI-driven insights

Conclusion
Incorporating Generative AI into your Enterprise Strategy Intelligence journey empowers organizations to harness the power of data-driven insights and make strategic decisions with greater confidence. By aligning your strategic goals, leveraging relevant data sources, and developing accurate Generative AI models, you can unlock new dimensions of enterprise strategy intelligence. Embrace this technology mindfully, while addressing ethical considerations, to stay at the forefront of innovation and drive sustainable success in today's dynamic business landscape.

Starting Your Journey: Using Generative AI to Define Your Strategic Goals

Introduction:
In today's fast-paced and highly competitive business landscape, defining clear strategic goals is essential for success. However, determining the right goals and strategies can be a complex and challenging process. Fortunately, advancements in technology, particularly in the field of generative artificial intelligence (AI), provide new opportunities to enhance strategic planning. Generative AI systems, such as OpenAI's GPT-3.5, can assist businesses in generating creative and innovative ideas, aiding in the formulation of effective strategic goals. This article explores how you can harness the power of generative AI to kickstart your journey towards defining strategic goals.

Starting Your Journey: Using Generative AI to Define Your Strategic Goals

Introduction:
In today's fast-paced and highly competitive business landscape, defining clear strategic goals is essential for success. However, determining the right goals and strategies can be a complex and challenging process. Fortunately, advancements in technology, particularly in the field of generative artificial intelligence (AI), provide new opportunities to enhance strategic planning. Generative AI systems, such as OpenAI's GPT-3.5, can assist businesses in generating creative and innovative ideas, aiding in the formulation of effective strategic goals. This article explores how you can harness the power of generative AI to kickstart your journey towards defining strategic goals.

1. Understanding Generative AI:
Generative AI refers to a subset of artificial intelligence that focuses on creating original and coherent content, such as text, images, or music. These AI systems are trained on vast amounts of data, enabling them to learn patterns, context, and language nuances. Generative AI models can then generate new content based on the patterns and information they have learned.

2. Leveraging Generative AI for Strategic Goal Definition:

a. Idea Generation: One of the most valuable applications of generative AI in strategic planning is idea generation. Traditional brainstorming sessions may be limited by the team's knowledge and experiences. Generative AI can overcome these limitations by offering fresh perspectives and generating ideas based on a vast corpus of information. By feeding the AI system with relevant data about your industry, market trends, and organizational objectives, you can receive a diverse range of creative ideas that can serve as a starting point for defining your strategic goals.

b. Exploring Possibilities: Generative AI can assist in exploring various possibilities and potential scenarios for your strategic goals. By inputting different parameters, objectives, and constraints into the AI model, you can generate multiple outcomes and assess their feasibility and potential outcomes. This approach helps you evaluate different strategic paths and make informed decisions.

c. Enhancing Decision-Making: Strategic decision-making often involves analyzing complex data and considering numerous variables. Generative AI can help streamline this process by providing insights and suggestions based on the input data and desired outcomes. By leveraging the AI system's ability to analyze patterns and trends, you can gain valuable insights that can influence your strategic goals.

3. Best Practices for Using Generative AI in Strategic Goal Definition:

a. Define Clear Inputs: To ensure meaningful and relevant outputs, it is crucial to provide clear inputs to the generative AI system. Clearly articulate your industry context, market dynamics, and organizational objectives when training the AI model. By doing so, you can enhance the system's understanding and generate more useful ideas.

b. Evaluate Outputs Critically: While generative AI can provide valuable insights, it is essential to evaluate the generated outputs critically. Not all ideas or suggestions may be feasible or align with your organizational goals. Human judgment and domain expertise are still essential to validate and refine the outputs generated by the AI system.

c. Iterative Approach: Strategic planning is an iterative process, and generative AI can aid in this iterative cycle. Continuously refine and update your AI model by incorporating feedback, analyzing real-time data, and adjusting parameters. This approach ensures that the AI system becomes increasingly tailored to your specific needs and goals.

Conclusion:

Incorporating generative AI into the strategic planning process can be a game-changer for organizations seeking to define their goals. By leveraging the power of AI to generate ideas, explore possibilities, and enhance decision-making, businesses can gain a competitive edge in today's dynamic marketplace. However, it is crucial to remember that generative AI is a tool, and human judgment remains indispensable in validating and implementing the generated outputs. By combining the power of generative AI with human expertise, organizations can embark on a transformative journey to define their strategic goals and drive future success.

Starting Your Journey: Using Generative AI to Identify Relevant Data Sources

Introduction:
Generative artificial intelligence (AI) has revolutionized various aspects of business, including strategic planning and decision-making. When embarking on a journey to leverage generative AI, one critical step is identifying relevant data sources. The quality and relevance of the data used to train the AI model greatly influence the accuracy and effectiveness of the generated outputs. In this section, we will explore the process of identifying relevant data sources to maximize the potential of generative AI in defining your strategic goals.

1. Defining Data Requirements:
Before diving into data sources, it's important to clearly define the data requirements for your generative AI model. Consider the specific aspects of your strategic planning process that you want the AI model to assist with. These could include market research, competitor analysis, customer insights, industry trends, financial data, and more. By identifying these requirements, you can narrow down the types of data sources needed for training the AI model.

2. Internal Data Sources:
a. Historical Data: Start by exploring your organization's internal data sources, such as databases, CRM systems, sales records, customer feedback, and operational data. This data provides insights into your company's past performance, customer behavior, and internal processes. Analyzing historical data can help identify patterns, trends, and potential areas for improvement, which can inform your strategic goals.

b. Financial Reports: Financial reports, including income statements, balance sheets, and cash flow statements, offer valuable insights into your company's financial health, profitability, and growth potential. By leveraging this data, generative AI can generate projections, financial models, and strategies aligned with your goals.

c. Employee Feedback and Surveys: Internal feedback from employees, such as surveys and engagement data, can provide valuable insights into the strengths and weaknesses of your organization. These inputs can help identify areas for improvement, innovation, and talent development, which can shape your strategic goals.

3. External Data Sources:
a. Market Research and Industry Reports: Utilize market research reports, industry publications, and relevant studies to gain a comprehensive understanding of your industry's landscape. These sources provide insights into market trends, consumer behavior, emerging technologies, and competitive analysis. Feeding this data into your generative AI model can help generate innovative strategies and identify opportunities for growth.

b. Social Media and Online Platforms: Social media platforms, discussion forums, and online communities are treasure troves of valuable data. Analyzing customer sentiments, product reviews, and online discussions can provide real-time insights into consumer preferences, industry buzz, and emerging trends. Incorporating this data into your AI model can generate ideas and strategies that resonate with your target audience.

c. Industry Blogs and Thought Leadership: Industry-specific blogs, thought leadership articles, and expert opinions provide deep insights into best practices, industry challenges, and emerging strategies. These sources can help you stay informed about the latest developments in your industry and generate innovative ideas for your strategic goals.

4. Data Quality and Pre-processing:
When identifying relevant data sources, it's crucial to prioritize data quality. Ensure that the data is accurate, up-to-date, and relevant to your specific needs. Additionally, consider pre-processing the data to remove noise, clean inconsistencies, and normalize formats. High-quality and well-structured data will enhance the performance and accuracy of your generative AI model.

Conclusion:
Identifying relevant data sources is a crucial step in harnessing the power of generative AI to define your strategic goals. By leveraging a combination of internal and external data sources, you can gain valuable insights into your organization's past performance, market dynamics, customer behavior, and industry trends. This data-driven approach, coupled with generative AI, empowers businesses to generate innovative strategies, make informed decisions, and stay ahead in a rapidly evolving business landscape. Remember, the success of your generative AI model relies heavily on the quality and relevance of the data sources you choose to incorporate.

Starting Your Journey: Building Generative AI Models

Introduction:
Generative artificial intelligence (AI) has emerged as a powerful tool that can create original and creative content, such as text, images, and music. Building your own generative AI model allows you to harness the potential of AI and unleash its creativity to generate unique outputs. In this section, we will explore the steps involved in building a generative AI model to kickstart your journey towards leveraging this technology for various applications.

1. Define the Problem and Objective:
Before diving into building a generative AI model, it is essential to clearly define the problem you want to solve and the specific objective you aim to achieve. For example, you may want to generate realistic human-like text, create unique images, compose music, or even develop conversational agents. Defining the problem and objective provides a clear direction for the model development process.

2. Collect and Preprocess Data:
Data forms the foundation for training a generative AI model. Depending on the objective, collect a substantial amount of relevant and high-quality data. For text generation, gather a diverse range of text documents, while for image generation, accumulate a large dataset of images. Ensure that the data is representative of the desired output and covers a wide range of styles and variations.

Preprocess the data by cleaning it, removing any inconsistencies or noise, and normalizing the format. This step is crucial to ensure the quality and effectiveness of the generative AI model.

3. Choose a Generative AI Architecture:
Selecting an appropriate generative AI architecture is crucial for achieving your desired outcomes. Several popular architectures include:

a. Recurrent Neural Networks (RNN): RNNs are well-suited for sequence generation tasks, such as text or music generation. Long Short-Term Memory (LSTM) and Gated Recurrent Units (GRU) are commonly used RNN variants that capture temporal dependencies in the data.

b. Variational Autoencoders (VAE): VAEs are useful for generating new data samples by learning a compact representation (latent space) of the input data. This architecture is widely used in image generation tasks.

c. Generative Adversarial Networks (GAN): GANs consist of two neural networks: a generator network that produces synthetic samples and a discriminator network that evaluates the authenticity of the generated samples. GANs are popular for generating realistic images and have achieved remarkable success in this domain.

Choose the architecture that aligns with your objective and explore existing research and implementations to understand their strengths and limitations.

4. Train the Generative AI Model:
Training the generative AI model involves feeding the preprocessed data into the chosen architecture. The model learns the underlying patterns and characteristics of the data to generate new content. This process typically requires significant computational resources and time, especially for complex models and large datasets. Training iterations involve adjusting the model's parameters, optimizing loss functions, and fine-tuning the network.

5. Evaluate and Refine the Model:
After training, evaluate the performance of your generative AI model. Use various metrics and qualitative assessments to measure the quality of the generated outputs. If the results are unsatisfactory, iterate on the model architecture, training parameters, or dataset to improve its performance. This iterative process of evaluation and refinement is crucial for enhancing the model's capabilities.

6. Generate New Content and Iterate:
Once you have a well-trained and validated generative AI model, it's time to generate new content based on the trained model. Experiment with different inputs, conditions, or styles to explore the model's creative potential. Iterate on the generated outputs, fine-tuning the model based on user feedback and domain-specific requirements.

Conclusion:
Building your own generative AI models empowers you to unlock the creative potential of AI technology. By following the steps outlined above, you can embark on a journey to build powerful generative AI models capable of generating unique and creative outputs. Remember that building generative AI models requires a deep understanding of the problem, careful data collection and preprocessing, selecting appropriate architectures, diligent training, evaluation, and iterative

refinement. With perseverance and an exploratory mindset, you can leverage generative AI models to revolutionize various domains and unlock new possibilities.

Starting Your Journey: Using Generative AI to Generate Insights and Validate

Introduction:
Generative artificial intelligence (AI) has revolutionized the way we generate insights and validate hypotheses in various domains. By leveraging the power of generative AI models, businesses and researchers can generate valuable insights, validate hypotheses, and explore new possibilities. In this section, we will explore how you can embark on a journey to use generative AI to generate insights and validate your ideas effectively.

1. Understanding Generative AI's Role in Insight Generation:
Generative AI models are capable of learning patterns, context, and nuances from large datasets. These models can generate new content that is coherent and aligned with the input data. When applied to insight generation, generative AI can assist in exploring and identifying patterns, trends, and relationships that may not be immediately apparent to humans. By analyzing and generating content based on existing data, generative AI can reveal hidden insights and novel perspectives.

2. Defining the Problem and Objectives:
To effectively use generative AI for insight generation and validation, it is essential to clearly define the problem you aim to address and the objectives you want to achieve. Consider the domain and specific questions you want to explore. Are you looking to understand customer preferences, optimize processes, or identify market trends? Clearly defining the problem will guide the model development process and ensure its alignment with your goals.

3. Collect and Prepare Relevant Data:
Data is the fuel that drives generative AI models. To generate meaningful insights and validate hypotheses, collect and prepare relevant data that is representative of the problem domain. This may include structured data, unstructured text, images, or any other form of data that aligns with your objectives. Cleanse and preprocess the data to remove noise, inconsistencies, and biases. Quality data ensures the accuracy and reliability of the insights generated by the AI model.

4. Choose the Appropriate Generative AI Model:
Selecting the right generative AI model depends on the nature of your data and the objectives you want to achieve. Some commonly used models include:

a. Variational Autoencoders (VAE): VAEs are well-suited for generating new data samples based on a learned latent space representation. They are commonly used in image generation and can provide valuable insights into data distributions and variations.

b. Transformer-based Models: Transformer-based models, such as OpenAI's GPT, are effective for generating text-based insights. They can analyze and generate coherent text that aligns with the provided input data. These models excel in tasks such as summarization, question-answering, and generating creative narratives.

c. Deep Convolutional Generative Adversarial Networks (DCGAN): DCGANs are popular for generating realistic images. They consist of a generator and a discriminator network that work in tandem to generate high-quality images that resemble the training data. DCGANs are ideal for validating visual concepts and generating visually appealing insights.

Choose the model that best suits your problem domain and aligns with the type of insights you aim to generate.

5. Train and Fine-tune the Model:

Train the generative AI model using the prepared data. The training process involves adjusting the model's parameters, optimizing loss functions, and iteratively improving the model's performance. The model learns from the patterns in the data and generates insights based on the learned information. Fine-tuning the model may be necessary to improve the quality and relevance of the generated insights.

6. Generate Insights and Validate Hypotheses:

Once the generative AI model is trained, use it to generate insights and validate hypotheses. Provide the model with specific input conditions or constraints to guide the generated outputs. Analyze the generated insights and evaluate their relevance, coherence, and alignment with the problem domain. These insights can help validate hypotheses, identify patterns, and gain a deeper understanding of the underlying data.

7. Iterative Refinement and Validation:

Generating insights and validating hypotheses through generative AI is an iterative process. Continuously refine the model, fine-tune parameters, and update the training data to improve the quality and relevance of the insights generated. Validate the insights against ground truth data or expert knowledge to ensure their accuracy and reliability.

Conclusion:

Generative AI provides a powerful tool for generating insights and validating hypotheses in various domains. By following the steps outlined above, you can embark on a journey to leverage generative AI models to uncover hidden patterns, explore new possibilities, and validate your ideas effectively. Remember to define the problem, collect relevant data, choose the appropriate generative AI model, train and fine-tune the model, generate insights, and validate them iteratively. With generative AI as your ally, you can unlock valuable insights and gain a competitive edge in your field of interest.

Starting Your Journey: Integrating Generative AI for Enterprise Strategy Intelligence

Introduction:

In the fast-paced and complex business landscape, organizations are increasingly turning to generative artificial intelligence (AI) to gain a competitive edge. One of the key areas where generative AI can make a significant impact is in integrating enterprise strategy intelligence into decision-making processes. By harnessing the power of generative AI, organizations can generate valuable insights, optimize strategic planning, and make informed decisions. In this section, we will explore how you can start your journey of integrating generative AI into your enterprise decision-making processes.

1. Define Your Strategic Objectives:

To effectively integrate generative AI into decision-making processes, it is crucial to define your strategic objectives clearly. Consider the key areas where you want to leverage AI-generated insights to drive your decision-making. This could include market analysis, competitor intelligence, customer segmentation, product development, resource allocation, or any other strategic aspect of your business. Identifying specific objectives ensures that the generative AI model is trained and aligned with your strategic needs.

2. Gather and Preprocess Relevant Data:

Generative AI models require high-quality and relevant data for training. Gather and preprocess the data that will serve as the foundation for generating strategic intelligence. This data can be derived from internal sources, such as operational data, customer records, sales data, or external sources like market research reports, industry data, and public datasets. Cleanse, transform, and normalize the data to ensure consistency and accuracy.

3. Choose the Appropriate Generative AI Model:

Select the generative AI model that best suits your strategic intelligence needs. Depending on the nature of the data and objectives, you may consider different models such as:

a. Natural Language Processing (NLP) Models: NLP models like OpenAI's GPT can generate text-based insights, such as summaries, market trend analysis, or competitive analysis. These models excel in processing and generating coherent text based on the provided input.

b. Deep Neural Networks (DNN) Models: DNN models can be utilized for tasks like demand forecasting, resource optimization, or predictive analytics. These models can analyze patterns, correlations, and historical data to generate insights and predictions that aid decision-making.

c. Generative Adversarial Networks (GANs): GANs are useful for generating visual representations, such as images, product designs, or marketing materials. These models can assist in creative decision-making and visualizing strategic concepts.

Choose the model that aligns with your strategic objectives and supports the type of intelligence you wish to generate.

4. Train and Validate the Generative AI Model:
Train the generative AI model using the preprocessed data. The training process involves feeding the data into the model, optimizing parameters, and iteratively refining the model's performance. It is crucial to validate the model's outputs against ground truth data, expert knowledge, or human feedback to ensure the generated intelligence is accurate, reliable, and aligned with your strategic objectives.

5. Integrate Generative AI into Decision-Making Processes:
Once the generative AI model is trained and validated, it's time to integrate it into your decision-making processes. Develop a framework that incorporates the generated intelligence into strategic planning, resource allocation, risk assessment, and performance evaluation. Collaborate with domain experts, data scientists, and decision-makers to ensure that the generated insights are effectively utilized and inform the decision-making process.

6. Monitor and Refine the Integration:
Regularly monitor the performance of the generative AI integration and collect feedback from stakeholders. Assess the impact of the generated intelligence on decision-making, and refine the model or data inputs if necessary. Continuously iterate and improve the integration to enhance the accuracy, relevance, and applicability of the generated insights.

Conclusion:
Integrating generative AI into decision-making processes empowers organizations to leverage strategic intelligence and make informed decisions. By defining strategic objectives, gathering relevant data, selecting appropriate generative AI models, training and validating the models, and integrating the generated intelligence into decision-making processes, organizations can enhance their strategic planning, optimize resource allocation, and gain a competitive advantage. Embrace generative AI as a strategic ally and embark on a journey to transform your enterprise decision-making through intelligent insights.

Starting Your Journey: Leveraging Generative AI to Monitor, Evaluate, and Adapt

Introduction:

In today's rapidly changing business landscape, organizations must continuously monitor, evaluate, and adapt their strategies to stay competitive. Generative artificial intelligence (AI) presents a powerful tool to assist in this process. By leveraging generative AI models, organizations can monitor market trends, evaluate performance, and adapt strategies based on valuable insights. In this section, we will explore how you can start your journey using generative AI to effectively monitor, evaluate, and adapt in the dynamic business environment.

1. Define Monitoring, Evaluation, and Adaptation Objectives:

To begin, clearly define your monitoring, evaluation, and adaptation objectives. Determine what aspects of your business you want to monitor and evaluate, such as customer behavior, market trends, operational performance, or competitive landscape. Identify the key metrics and indicators that will help you gauge success and performance in these areas. Defining specific objectives ensures that the generative AI models are trained to deliver the insights needed to monitor, evaluate, and adapt effectively.

2. Gather Relevant Data:

To enable generative AI models to provide meaningful insights, gather relevant data from various sources. This may include internal data from sales, marketing, operations, and finance, as well as external data from market research reports, industry trends, social media, or customer feedback. The data should cover a wide range of factors that influence your business and align with your monitoring and evaluation objectives.

3. Preprocess and Prepare Data:

Preprocessing and preparing the data are critical steps to ensure the accuracy and reliability of the generated insights. Cleanse the data by removing duplicates, outliers, and inconsistencies. Normalize and standardize the data to make it compatible for analysis by the generative AI models. Data preprocessing helps enhance the quality of the generated insights and enables effective monitoring and evaluation.

4. Select the Appropriate Generative AI Models:

Choosing the right generative AI models is essential to effectively monitor, evaluate, and adapt. Consider the nature of the data and the objectives you want to achieve. Some commonly used generative AI models include:

a. Time-Series Analysis Models: Time-series models, such as recurrent neural networks (RNNs) or long short-term memory (LSTM) networks, are suitable for monitoring and predicting trends over time. These models can analyze historical data and generate forecasts to support strategic decision-making.

b. Clustering and Classification Models: Clustering and classification models, like k-means or support vector machines (SVM), can help identify patterns and segment data into meaningful categories. These models are valuable for customer segmentation, market analysis, and identifying anomalies.

c. Reinforcement Learning Models: Reinforcement learning models enable adaptive decision-making by learning from feedback and optimizing strategies based on rewards. These models are useful for adaptive resource allocation, pricing optimization, and dynamic decision-making.

Select the generative AI models that align with your monitoring, evaluation, and adaptation objectives to generate insightful outputs.

5. Train and Validate the Generative AI Models:
Train the selected generative AI models using the preprocessed data. This training process involves feeding the data into the models, optimizing parameters, and iteratively refining the models' performance. Validate the outputs generated by the models using ground truth data, expert feedback, or domain knowledge to ensure the accuracy and reliability of the insights.

6. Monitor, Evaluate, and Adapt:
Once the generative AI models are trained and validated, it's time to put them into action. Monitor the relevant metrics and data points identified in your objectives. Analyze the outputs generated by the generative AI models to evaluate performance, identify trends, and detect anomalies. Regularly assess the insights provided by the models to inform decision-making and adapt strategies accordingly. Continuously monitor, evaluate, and adapt based on the insights generated by the generative AI models to stay responsive and agile in the ever-changing business landscape.

7. Refine and Improve:
As you progress on your journey, continually refine and improve your generative AI models and processes. Incorporate feedback from stakeholders, evaluate the impact of the insights on decision-making, and refine the models' parameters or data inputs if necessary. Embrace a culture of learning and adaptability to continuously enhance the effectiveness and relevance of the generative AI-driven monitoring, evaluation, and adaptation processes.

Conclusion:
Leveraging generative AI for monitoring, evaluation, and adaptation empowers organizations to make data-driven decisions in a dynamic business environment. By defining objectives, gathering relevant data, selecting appropriate generative AI models, training and validating the models, and implementing a robust monitoring, evaluation, and adaptation framework, organizations can gain valuable insights and adapt strategies effectively. Embrace generative AI as a powerful tool in your journey towards agility, responsiveness, and sustainable growth.

Figure 17

Chapter 16: Preparing Your Organization for Enterprise Strategy Intelligence Deployment: Driving Enterprise Agility Across Functional Teams

Introduction:

In today's rapidly evolving business landscape, organizations must embrace agility to stay ahead of the competition and respond effectively to market dynamics. Enterprise Strategy Intelligence (ESI) offers a powerful framework that leverages data-driven insights to inform strategic decision-making. By deploying ESI across all functional teams, organizations can foster a culture of agility and empower teams to make informed decisions aligned with the broader strategic objectives. In this section, we will explore how organizations can prepare themselves to deploy ESI across functional teams, ultimately driving enterprise agility.

1. **Establish a Clear Strategic Vision:**
   Before deploying ESI, it is crucial to establish a clear strategic vision that outlines the organization's goals, objectives, and desired outcomes. This vision serves as the guiding compass for the deployment process and ensures alignment across all functional teams. Clearly communicate the strategic vision to all stakeholders and emphasize the importance of agility in achieving organizational success.

2. Develop a Data-Driven Culture:

To effectively deploy ESI, organizations need to foster a data-driven culture where information and insights play a central role in decision-making. Educate and train employees on the importance of leveraging data and analytics in their respective functional areas. Encourage the use of data-backed insights to support decision-making processes and emphasize the benefits of agility in responding to market changes.

3. Build Cross-Functional Collaboration:

ESI deployment requires collaboration and integration across functional teams. Break down silos and foster collaboration between different departments, such as marketing, finance, operations, and sales. Encourage regular communication, knowledge sharing, and cross-functional projects to promote a holistic understanding of the organization's strategic objectives. Cross-functional collaboration enables teams to leverage ESI insights collectively and make informed decisions that align with broader strategic goals.

4. Establish Data Infrastructure and Integration:

To enable effective ESI deployment, organizations must establish a robust data infrastructure that supports data collection, integration, and analysis across functional teams. Implement data management systems that allow for efficient data gathering, storage, and processing. Integrate various data sources, both internal and external, to create a comprehensive view of the business landscape. Ensure data security and compliance with relevant regulations to maintain the integrity and confidentiality of sensitive information.

5. Implement ESI Tools and Technologies:

Invest in appropriate ESI tools and technologies that facilitate data analysis, visualization, and reporting. These tools should be accessible to all functional teams, enabling them to leverage ESI insights in their day-to-day activities. Choose user-friendly platforms that support data exploration, predictive modeling, scenario planning, and performance tracking. Provide training and support to ensure teams can effectively utilize these tools to derive actionable insights.

6. Define Key Performance Indicators (KPIs):

To drive enterprise agility, establish relevant Key Performance Indicators (KPIs) that align with the strategic objectives of the organization. Develop KPIs that are measurable, specific, and time-bound. These KPIs will serve as benchmarks for tracking performance and evaluating the impact of ESI deployment. Regularly assess and refine the KPIs to ensure they remain aligned with the evolving business landscape.

7. Continuously Monitor, Evaluate, and Adapt:

ESI deployment is an ongoing process that requires continuous monitoring, evaluation, and adaptation. Regularly review the ESI insights generated by functional teams and assess their impact on decision-making and organizational performance. Encourage a feedback loop where teams provide insights and recommendations based on ESI data, allowing for iterative refinement of strategies and approaches.

Conclusion:

Deploying Enterprise Strategy Intelligence (ESI) across functional teams is a crucial step towards driving enterprise agility. By establishing a clear strategic vision, fostering a data-driven culture, promoting cross-functional collaboration, implementing robust data infrastructure and ESI tools, defining relevant KPIs, and continuously monitoring and adapting, organizations can unlock the full potential of ESI. Embrace ESI as a strategic enabler and empower your functional teams to make informed decisions aligned with organizational objectives, ultimately driving enterprise agility and staying ahead in the competitive landscape.

Unleashing the Power of Enterprise Strategy Intelligence: Preparing Organizations for Consistent and Agile Execution

Introduction:

In today's fast-paced business environment, organizations must not only formulate robust strategies but also ensure their seamless execution across teams. To achieve this, organizations need to embrace the concept of Enterprise Strategy Intelligence (ESI), a holistic approach that combines real-time data, analytics, and effective communication to align teams and enable consistent execution at a rapid pace. In this section, we will explore key steps to prepare an organization to execute at the same speed and deliver consistency across teams using ESI.

1. Establish Clear Strategic Objectives:

To execute at the same speed, organizations must establish clear strategic objectives that align with their overall vision. These objectives should be specific, measurable, achievable, relevant, and time-bound (SMART). By setting clear goals, organizations can provide teams with a unified sense of purpose, allowing them to focus their efforts and work cohesively towards shared outcomes.

2. Foster a Culture of Collaboration:

Successful execution requires strong collaboration across teams and departments. Organizations should create an environment that encourages open communication, knowledge sharing, and cross-functional collaboration. This can be achieved through regular meetings, team-building exercises, and the use of collaboration tools that facilitate seamless information exchange.

3. Invest in Enterprise Strategy Intelligence Tools:

To enable consistent execution, organizations should invest in ESI tools that provide real-time insights and analytics. These tools can help track progress, identify bottlenecks, and measure key performance indicators (KPIs) across various teams and projects. By leveraging data-driven intelligence, organizations can make informed decisions and quickly adapt their strategies based on changing market conditions.

4. Implement Agile Project Management Practices:

Agile methodologies have gained popularity due to their ability to enhance team efficiency and adaptability. By adopting agile project management practices, organizations can break down complex projects into manageable tasks, set iterative goals, and promote continuous improvement. Agile practices facilitate faster decision-making, shorter feedback loops, and increased flexibility, allowing teams to respond promptly to evolving business needs.

5. Enable Effective Communication Channels:

Effective communication is the cornerstone of consistent execution. Organizations should establish clear and transparent communication channels that facilitate the exchange of information and ensure everyone is on the same page. This can include regular status updates, progress reports, team huddles, and the use of digital collaboration platforms. Moreover, leaders should encourage feedback and actively listen to team members' perspectives to foster a culture of trust and empowerment.

6. Provide Ongoing Training and Development:
To execute at the same speed, organizations must invest in the continuous training and development of their employees. This includes providing access to relevant resources, workshops, and training programs that enhance skills, knowledge, and awareness of market trends. By investing in their workforce, organizations can build a highly skilled and adaptable team that can execute strategies effectively.

7. Embrace a Data-Driven Decision-Making Culture:
ESI relies heavily on data and analytics to drive decision-making. Organizations should encourage a data-driven culture where decisions are based on objective insights rather than gut feelings. By leveraging data analytics tools, organizations can gain valuable insights into customer behavior, market trends, and operational performance, enabling them to make informed decisions and execute strategies with precision.

Conclusion:
In today's rapidly changing business landscape, organizations that can execute at the same speed and deliver consistency across teams will have a significant competitive advantage. By embracing the power of Enterprise Strategy Intelligence, organizations can align teams, harness real-time data, and foster a culture of collaboration and agility. Through clear strategic objectives, effective communication channels, and the use of ESI tools, organizations can position themselves for success in an environment that demands speed, adaptability, and consistent execution.

Scaling Agile: Driving Faster Solutions with Enterprise Strategy Intelligence

Introduction:
In an era of rapid technological advancements and dynamic market conditions, organizations must be able to scale their operations and adapt quickly to drive faster solutions. The application of Agile methodologies at the enterprise level, combined with the power of Enterprise Strategy Intelligence (ESI), can enable organizations to embrace flexibility, collaboration, and data-driven decision-making. In this section, we will explore the key steps to prepare an organization to scale and apply the Agile model at the enterprise level, leveraging ESI to drive faster and more effective solutions.

1. Develop an Agile Mindset:
Scaling Agile begins with cultivating an Agile mindset throughout the organization. This involves promoting a culture of openness, adaptability, and continuous improvement. Leaders must champion the Agile values and principles, emphasizing the importance of collaboration, customer-centricity, and embracing change. By fostering an Agile mindset, organizations can create an environment where teams are empowered to experiment, learn from failures, and iterate quickly to deliver innovative solutions.

2. Establish Agile Governance:
To effectively scale Agile across the enterprise, organizations need to establish an Agile

governance framework. This framework defines the rules, processes, and roles necessary for Agile implementation at a larger scale. It includes establishing cross-functional teams, defining decision-making structures, and ensuring alignment with strategic objectives. Agile governance provides the necessary structure to ensure consistency, transparency, and accountability across teams while maintaining the flexibility Agile methodologies offer.

3. Align Agile with Enterprise Strategy:

Successful scaling of Agile requires a strong alignment between Agile initiatives and the overall enterprise strategy. Organizations should ensure that Agile teams and projects are directly linked to strategic goals, enabling them to drive meaningful outcomes aligned with the organization's vision. This alignment helps prioritize efforts, allocate resources effectively, and ensure that Agile initiatives contribute to the organization's long-term success.

4. Leverage Enterprise Strategy Intelligence:

Enterprise Strategy Intelligence (ESI) plays a vital role in driving faster solutions in an Agile environment. ESI involves leveraging real-time data, advanced analytics, and intelligent insights to inform decision-making and strategy execution. By utilizing ESI tools and platforms, organizations can gain a comprehensive view of their operations, customer behavior, and market dynamics. This enables data-driven decision-making, facilitates rapid identification of opportunities and risks, and empowers teams to respond quickly and effectively.

5. Encourage Cross-Functional Collaboration:

Effective scaling of Agile requires breaking down silos and promoting cross-functional collaboration. Agile teams should consist of members from different departments and disciplines to ensure diverse perspectives and holistic problem-solving. Organizations should invest in fostering a collaborative culture, breaking communication barriers, and providing tools that facilitate seamless collaboration, such as digital workspaces and virtual communication platforms. Cross-functional collaboration enhances innovation, accelerates problem-solving, and drives faster solutions.

6. Prioritize Continuous Learning and Improvement:

Scaling Agile at the enterprise level necessitates a commitment to continuous learning and improvement. Organizations should establish mechanisms for sharing best practices, conducting retrospectives, and capturing lessons learned. Regular feedback loops and retrospectives enable teams to reflect on their performance, identify areas for improvement, and implement iterative changes. By embracing a culture of continuous learning, organizations can enhance their Agile practices, streamline processes, and drive faster and more efficient solutions.

7. Invest in Agile Training and Coaching:

To effectively scale Agile, organizations should invest in training and coaching programs to equip employees with the necessary skills and knowledge. Agile training should encompass not only the principles and practices but also the mindset and values that underpin Agile methodologies. Additionally, organizations should provide ongoing coaching and support to Agile teams, ensuring they have the guidance and resources needed to succeed in their Agile journey.

Conclusion:

Scaling Agile at the enterprise level requires a strategic approach that embraces an Agile mindset, aligns with organizational strategy, and leverages the power of Enterprise Strategy Intelligence. By fostering collaboration, establishing Agile governance, and investing in training, organizations can create an environment that enables faster solutions, iterative improvements, and the ability to adapt swiftly to changing market demands. By harnessing the benefits of Agile methodologies and ESI, organizations can position themselves for success in an increasingly dynamic and competitive business landscape.

Unlocking the Power of Speed and Scale: Preparing Your Organization for Data-to-Action with Enterprise Strategy Intelligence

Introduction

In today's data-driven world, organizations that can swiftly transform data into actionable insights gain a significant competitive advantage. To harness the power of data effectively, companies need to adopt a data-led approach, where decision-making is driven by insights derived from robust data analysis. This article explores the concept of enterprise strategy intelligence and outlines key steps for organizations to prepare themselves for rapid data-to-action capabilities with speed and scale.

Understanding Enterprise Strategy Intelligence

Enterprise Strategy Intelligence (ESI) refers to the systematic and comprehensive use of data and analytics to inform strategic decision-making within an organization. It involves the integration of various data sources, advanced analytics techniques, and agile processes to generate actionable insights that guide business strategies, operations, and innovation initiatives. ESI empowers organizations to leverage data as a strategic asset and transform it into meaningful actions quickly and efficiently.

Key Steps to Prepare Your Organization for Data-to-Action with Speed and Scale

1. **Cultivate a Data-Driven Culture:** Building a data-led company starts with fostering a culture that values and embraces data. Encourage employees at all levels to be curious, experiment with data, and make decisions based on evidence rather than intuition alone. Establish data literacy programs and training initiatives to equip your workforce with the skills needed to interpret and analyze data effectively.

2. **Establish a Robust Data Infrastructure:** To enable swift data-to-action capabilities, organizations must invest in a robust data infrastructure. This involves ensuring data quality, integration, and accessibility. Implement data governance practices to maintain data integrity, and leverage modern data management technologies to handle large volumes of data efficiently. Cloud-based solutions can provide scalability and flexibility for handling data at scale.

3. **Embrace Advanced Analytics:** Traditional business intelligence approaches are no longer sufficient for driving data-to-action. Embrace advanced analytics techniques such as predictive modeling, machine learning, and natural language processing to extract deeper insights from your data. Deploy data science teams or collaborate with external partners to leverage their expertise in developing sophisticated models and algorithms.

4. **Enable Real-Time Data Processing:** In today's fast-paced business environment, real-time data processing is crucial. Implement technologies and platforms that enable

real-time data ingestion, processing, and analysis. This allows your organization to respond swiftly to emerging trends, customer demands, and market shifts. Consider adopting stream processing frameworks and event-driven architectures to handle data in motion effectively.

5. **Foster Collaboration Across Departments:** Siloed data and fragmented decision-making processes hinder the speed and scale of data-to-action capabilities. Encourage cross-functional collaboration by breaking down data silos and promoting information sharing across departments. Implement collaborative tools and platforms that facilitate knowledge exchange and enable teams to work together seamlessly.

6. **Establish Agile Decision-Making Processes**: Agility is a critical component of leveraging data for quick action. Implement agile decision-making processes that enable rapid experimentation, fail-fast mentality, and continuous learning. Break down complex problems into smaller, manageable tasks and iterate quickly based on feedback and data-driven insights. Embrace agile project management methodologies such as Scrum or Kanban to drive faster and more effective decision-making.

7. **Invest in Data Visualization and Reporting:** Effective data communication is vital for driving action. Invest in intuitive data visualization tools that allow stakeholders to interact with data and gain insights effortlessly. Establish robust reporting mechanisms to disseminate key findings across the organization promptly. Visual representations of data enable faster understanding and facilitate decision-making at all levels.

Conclusion

In an era defined by data, organizations must evolve into data-led entities that can rapidly transform insights into action. By adopting enterprise strategy intelligence and embracing a data-driven culture, organizations can leverage their data assets to drive strategic decision-making. Implementing a robust data infrastructure, embracing advanced analytics, enabling real-time data processing, fostering collaboration, establishing agile decision-making processes, and investing in data visualization are crucial steps toward becoming a data-led company. By following these steps, organizations can unleash the power of speed and scale, gaining a competitive edge in today's dynamic business landscape.

Unleashing the Potential of Enterprise Strategy Intelligence: Seamlessly Aggregating and Moving Data Across the Organization

Introduction

In the age of data-driven decision-making, organizations are recognizing the importance of harnessing the full potential of their data assets. Enterprise Strategy Intelligence (ESI) provides a powerful framework for organizations to leverage data effectively in driving strategic decision-making. One critical aspect of implementing ESI is the seamless aggregation and movement of data throughout the enterprise. In this section, we will explore the key considerations and steps organizations can take to ensure a smooth flow of data across their organization, enabling the full realization of ESI.

1. Establish a Data Governance Framework

To ensure the seamless movement of data, it is essential to establish a robust data governance framework. This framework should define data ownership, establish data standards, and specify data management policies and procedures. Clear guidelines on data classification, access controls, data quality, and data privacy will help ensure that data flows smoothly and securely across the organization.

2. Implement Data Integration and ETL Processes

Efficient data integration is a cornerstone of seamless data movement. Organizations should invest in technologies and tools that facilitate the extraction, transformation, and loading (ETL) of data from various sources into a central data repository. Implementing robust ETL processes ensures that data is cleansed, standardized, and harmonized before being made available for analysis and decision-making.

3. Leverage Data Warehousing and Data Lakes

Data warehousing and data lakes serve as central repositories for storing and organizing large volumes of structured and unstructured data. Data warehouses are ideal for structured data that requires high levels of data consistency, while data lakes accommodate diverse data types and formats. Leveraging these storage solutions ensures that data from different sources can be easily aggregated, enabling comprehensive analysis and insights generation.

4. Embrace Data Integration Technologies

To facilitate seamless data movement, organizations should leverage advanced data integration technologies. These technologies enable the integration of data from disparate sources, such as databases, cloud platforms, and external data providers. Data integration tools, extract-transform-load (ETL) software, and application programming interfaces (APIs) can automate the movement of data, ensuring a consistent and reliable flow of information across systems.

5. Implement Data Virtualization

Data virtualization is a technique that allows organizations to access and combine data from multiple sources without physically moving or replicating it. By creating a virtual layer that abstracts the underlying data sources, data virtualization enables real-time access to integrated data views. This approach reduces data duplication, enhances data consistency, and enables faster data retrieval and analysis.

6. Enable Data Collaboration and Sharing

Facilitating data collaboration and sharing is crucial for seamless data movement within an organization. Implementing collaborative platforms, data sharing portals, and knowledge-sharing tools encourages cross-functional teams to access and contribute to data repositories. This fosters a culture of collaboration, accelerates decision-making, and ensures that relevant stakeholders have timely access to the data they need.

7. Implement Data Governance and Security Measures

As data flows across the organization, it is essential to implement robust data governance and security measures. This includes establishing access controls, encryption protocols, and monitoring mechanisms to safeguard data integrity and protect against unauthorized access. Regular audits and compliance checks should be conducted to ensure adherence to data governance policies and industry regulations.

8. Embrace Automation and Orchestration

To enhance the efficiency of data movement, organizations should embrace automation and orchestration capabilities. Workflow automation tools and data orchestration platforms can streamline data movement processes, eliminating manual interventions and reducing the risk of errors. Automated data pipelines ensure that data is moved seamlessly, transformed, and made available to stakeholders in a timely manner.

Conclusion

To fully leverage the power of Enterprise Strategy Intelligence, organizations must ensure the seamless aggregation and movement of data across their enterprise. By establishing a strong data governance framework, implementing data integration processes, leveraging data warehousing and data lakes, embracing data integration technologies and data virtualization, enabling data collaboration and sharing, implementing robust data governance and security measures, and embracing automation and orchestration, organizations can unlock the full potential of their data assets. By seamlessly moving data, organizations can derive valuable insights, make informed decisions, and gain a competitive advantage in today's data-driven business landscape.

Moving Beyond Predictive Models: Unleashing the Power of Generative AI with Enterprise Strategy Intelligence

Introduction

In the realm of data-driven decision-making, predictive models have long been a valuable tool for organizations. However, as technology advances, organizations are now exploring new frontiers with generative artificial intelligence (AI). Generative AI enables organizations to move beyond predictions and venture into the realm of generating new and innovative possibilities. By leveraging generative AI within the framework of Enterprise Strategy Intelligence (ESI), organizations can unlock untapped potential for strategic decision-making. This article explores the concept of generative AI and outlines how organizations can prepare to harness its power within the context of ESI.

Understanding Generative AI

Generative AI refers to a class of AI algorithms and models that have the ability to generate new, original content, such as images, text, audio, or even entire scenarios. Unlike predictive models that analyze existing data to make predictions, generative AI models can create entirely new data that resembles the patterns and characteristics of the training data. This capability opens up new avenues for creativity, innovation, and strategic thinking within organizations.

Preparing for Generative AI within ESI

1. **Establish a Strong Data Foundation:** Generative AI models require a robust and diverse dataset for training. Organizations should focus on curating and aggregating high-quality data from various sources relevant to their business domain. A comprehensive and well-organized dataset serves as the foundation for training generative AI models effectively.

2. **Invest in Generative AI Expertise:** Developing expertise in generative AI is essential for organizations looking to leverage this technology. This may involve hiring data scientists, machine learning engineers, or partnering with external experts who specialize in generative AI. Organizations should invest in training and upskilling their teams to ensure they possess the necessary skills and knowledge to work with generative AI models.

3. **Define Clear Objectives and Use Cases:** Before implementing generative AI, organizations need to define clear objectives and identify specific use cases where

generative AI can add value. Whether it's generating creative content, simulating scenarios, or exploring alternative strategies, a focused approach will help guide the development and implementation of generative AI within the organization.

4. **Collect and Preprocess Data:** Generative AI models require large amounts of high-quality training data. Organizations should collect and preprocess data to ensure its cleanliness, relevance, and suitability for the intended use case. Data preprocessing techniques such as normalization, noise reduction, and feature extraction play a crucial role in preparing data for effective training of generative AI models.

5. **Develop and Train Generative AI Models:** Once the data is ready, organizations can start developing and training generative AI models. This involves selecting appropriate algorithms, architectures, and frameworks based on the use case. Experimentation and iteration are key to fine-tuning the models and achieving the desired outcomes.

6. **Embrace Ethical Considerations:** Generative AI brings with it ethical considerations, particularly when generating content that resembles real-world data. Organizations must be mindful of potential biases, misinformation, or unintended consequences that may arise from the use of generative AI. Implementing ethical guidelines and robust validation processes are necessary to ensure responsible and trustworthy use of generative AI.

7. **Integrate Generative AI into Decision-Making Processes:** To leverage the power of generative AI effectively, organizations should integrate it into their decision-making processes within the framework of ESI. By combining generative AI outputs with predictive models, real-time data, and human expertise, organizations can explore new possibilities, simulate scenarios, and generate innovative strategies.

8. **Continuously Evaluate and Refine:** Generative AI models require continuous evaluation and refinement to ensure their effectiveness and relevance. Organizations should establish mechanisms for monitoring the performance of generative AI models, gathering feedback from users, and incorporating improvements based on user insights and evolving business needs.

Conclusion

Generative AI represents a paradigm shift in strategic decision-making, allowing organizations to move beyond predictive models and explore new frontiers of creativity and innovation. By integrating generative AI within the framework of Enterprise Strategy Intelligence, organizations can leverage its power to generate new possibilities, simulate scenarios, and develop innovative strategies. By preparing their data, investing in expertise, defining clear objectives, developing and training models, considering ethical implications, integrating generative AI into decision-making processes, and continuously evaluating and refining, organizations can embrace the transformative potential of generative AI within their strategic initiatives.

Determining Success and Measuring the Successful Launch of Enterprise Strategy Intelligence with Generative AI

Introduction

In today's rapidly evolving business landscape, enterprises are constantly seeking innovative ways to gain a competitive edge and stay ahead of the curve. One area that has garnered significant attention is the integration of generative artificial intelligence (AI) into enterprise strategy intelligence. By leveraging the power of generative AI, organizations can unlock new possibilities for generating valuable insights, improving decision-making processes, and ultimately achieving success. However, determining the success of such initiatives and effectively measuring their impact require careful consideration. This article explores the key factors for evaluating the successful launch of enterprise strategy intelligence with generative AI.

Defining Enterprise Strategy Intelligence with Generative AI

Enterprise strategy intelligence refers to the systematic collection, analysis, and utilization of information to support strategic decision-making within an organization. It involves gathering data from various sources, transforming it into actionable insights, and aligning business strategies accordingly. Generative AI, on the other hand, involves the use of machine learning algorithms to generate novel and creative outputs, such as text, images, or even strategies, based on patterns and knowledge acquired from vast datasets.

Factors for Determining Success

1. **Clear Objectives:** The first step in determining the success of launching enterprise strategy intelligence with generative AI is to define clear objectives. These objectives should align with the organization's overall strategic goals and address specific pain points or challenges. Common objectives may include enhancing market intelligence, optimizing resource allocation, improving customer experience, or identifying new growth opportunities.

2. **Data Quality and Accessibility:** The success of generative AI relies heavily on the quality and accessibility of data. Enterprises must ensure that they have access to diverse and relevant datasets that can effectively train the AI models. The data should be accurate, up-to-date, and cover a wide range of variables to generate meaningful insights. Robust data governance practices and data infrastructure are crucial for success.

3. **AI Model Development:** Developing a robust AI model tailored to the specific needs of the enterprise is vital. This involves selecting appropriate generative AI algorithms, training the model using relevant data, and fine-tuning it to achieve optimal performance. The model should be capable of generating insights that align with the organization's strategic objectives and provide actionable recommendations

4. **Integration and Adoption:** Successful integration of generative AI into existing enterprise systems and processes is crucial for its effective utilization. The AI-powered insights should be seamlessly integrated into decision-making frameworks, enabling stakeholders to make informed choices. The organization must also focus on fostering a culture of AI adoption and provide the necessary training and support to ensure widespread acceptance and utilization.

Measuring Success

1. **Impact on Decision-Making:** One of the primary indicators of success is the impact of generative AI on decision-making processes. Assessing whether the AI-generated insights have led to more informed and effective decision-making can provide valuable insights into the success of the initiative. This can be measured through qualitative feedback from decision-makers and quantitatively by comparing key performance indicators (KPIs) before and after AI integration.

2. **Return on Investment (ROI):** Evaluating the financial impact of implementing generative AI is another crucial aspect of measuring success. ROI can be assessed by analyzing cost savings, revenue growth, and operational efficiency improvements resulting from AI-driven insights. Additionally, organizations can compare the costs associated

with developing and maintaining the AI infrastructure against the benefits derived from its implementation.

3. **User Satisfaction:** User satisfaction plays a vital role in the long-term success of enterprise strategy intelligence with generative AI. Gathering feedback from stakeholders, including executives, managers, and end-users, about their experience with the AI-generated insights can help gauge satisfaction levels. Surveys, interviews, and user feedback mechanisms can be used to collect this valuable information.

4. **Adaptability and Scalability:** The ability of the generative AI solution to adapt to evolving business needs and scale across different departments or business units is an important factor in measuring success. If the solution proves to be adaptable, scalable, and capable of addressing a wide range of strategic challenges, it signifies a successful implementation.

Conclusion

The successful launch of enterprise strategy intelligence with generative AI requires careful planning, robust AI model development, and effective integration and adoption. By defining clear objectives, ensuring data quality and accessibility, and measuring factors such as impact on decision-making, ROI, user satisfaction, and adaptability, organizations can evaluate the success of their AI initiatives. Ultimately, the successful utilization of generative AI in enterprise strategy intelligence has the potential to revolutionize decision-making processes, drive innovation, and enhance competitiveness in today's dynamic business environment.

risk assessment

Figure 18

Chapter 17: Readiness Assessments, Checklist, Scorecard, KPI for Generative AI for Enterprise Decision Intelligence

Assessment for Evaluating Organizational Readiness to Integrate Enterprise Strategy Intelligence

For each assessment criterion, the organization can be rated on a scale of 1 to 5, with 1 indicating a low readiness level and 5 indicating a high readiness level. The scores can be aggregated to provide an overall readiness score and identify areas that require attention and improvement. This assessment serves as a starting point to evaluate the organization's readiness to integrate enterprise strategy intelligence and can be customized based on specific requirements and context.

Readiness Assessment for Enterprise Strategy Intelligence with Generative AI

Item	Topic	Definition
1	**Leadership and Vision:**	• Is there a clear vision for integrating enterprise strategy intelligence? • Are senior leaders committed to driving intelligence initiatives? • Do leaders understand the potential impact of intelligence on strategic decision-making? • Are resources allocated to support intelligence integration efforts?
2	**Data Infrastructure and Governance:**	• Is there a robust data infrastructure in place to support intelligence initiatives? • Are data sources integrated and accessible across the organization? • Is there a data governance framework to ensure data quality and compliance? • Are privacy and security measures implemented to protect sensitive data?
3	**Analytical Capabilities:**	• Does the organization have skilled analysts and data scientists? • Are there analytics tools and technologies in place to support intelligence activities? • Are employees trained in data analysis and interpretation? • Is there a culture of data-driven decision-making within the organization?
4	**Strategic Alignment:**	• Is the organization's overall strategy clearly defined and communicated? • Are intelligence initiatives aligned with strategic goals and objectives? • Are there mechanisms in place to ensure alignment between intelligence and strategy? • Are key performance indicators (KPIs) identified to measure the impact of intelligence on strategy execution?

Item	Topic	Definition
5	**Stakeholder Engagement:**	• Are stakeholders involved in the design and implementation of intelligence initiatives? • Are there communication channels to keep stakeholders informed about intelligence activities? • Are stakeholders receptive to leveraging intelligence for decision-making? • Is there a feedback mechanism to capture stakeholder needs and expectations?
6	**Change Management and Culture:**	• Is there a change management strategy to support the integration of intelligence? • Is the organization open to embracing new ways of working and adopting intelligence practices? • Are there initiatives to foster a data-driven culture within the organization? • Are employees empowered and encouraged to use intelligence in their day-to-day activities?
7	**Performance Measurement and Continuous Improvement:**	• Are there mechanisms to measure and evaluate the effectiveness of intelligence integration? • Is performance regularly reviewed against defined metrics and targets? • Are there feedback loops to capture lessons learned and drive continuous improvement? • Are there initiatives in place to share best practices and knowledge across the organization?

Success Checklist: Determining the Successful Launch of Enterprise Strategy Intelligence with Generative AI

By considering and evaluating each of these checklist items, organizations can assess the success of their enterprise strategy intelligence initiative powered by generative AI. It allows them to identify areas of strength and areas that may require further attention or improvement, ultimately leading to enhanced decision-making processes and strategic outcomes.

Item	Topic	Definition
1	**Clearly Defined Objectives:**	Are the objectives of the enterprise strategy intelligence initiative clearly defined and aligned with the organization's overall strategic goals?
2	**Data Auality and Accessibility:**	Is there access to diverse and relevant datasets that can effectively train the generative AI models? Is the data accurate, up-to-date, and comprehensive enough to generate meaningful insights?
3	**AI Model Development:**	Has a robust generative AI model been developed, tailored to the specific needs of the organization? Has the model been trained using relevant data and fine-tuned for optimal performance?
4	**Integration And Adoption:**	Has the generative AI solution been seamlessly integrated into existing enterprise systems and processes? Is there widespread acceptance and utilization of the AI-generated insights among decision-makers and end-users?
5	**Impact On Decision-Making:**	Has the generative AI solution had a positive impact on decision-making processes? Are decision-makers able to make more informed and effective choices based on the AI-generated insights?
6	**Return On Investment (ROI):**	Has the implementation of generative AI resulted in measurable financial benefits? Has there been a positive impact on cost savings, revenue growth, and operational efficiency?
7	**User Satisfaction:**	Are stakeholders, including executives, managers, and end-users, satisfied with the AI-generated insights? Have their feedback and user experiences been collected and analyzed?
8	**Adaptability and Scalability:**	Is the generative AI solution adaptable to evolving business needs? Can it scale across different departments or business units? Is it capable of addressing a wide range of strategic challenges?

Item	Topic	Definition
9	**Ethical Considerations:**	Have ethical considerations been taken into account throughout the development and implementation of the generative AI solution? Has the organization ensured fairness, transparency, and accountability in the AI processes?
10	**Continuous Improvement:**	Is there a plan in place for continuous improvement of the generative AI solution? Are there mechanisms for monitoring and evaluating its performance over time and making necessary adjustments?
11	**Stakeholder Communication and Support:**	Has effective communication and support been provided to stakeholders throughout the launch and implementation process? Are there mechanisms for addressing any concerns or challenges that arise?

Scorecard for Enterprise Intelligence Initiatives

Each criterion can be assigned a score on a scale of 1 to 10, with 10 being the highest. The weightage indicates the importance of each criterion in the overall assessment. The scores can be aggregated to calculate an overall score for the enterprise intelligence initiatives, providing a quantitative measure of their effectiveness and progress. Regular review and adjustment of the scorecard may be necessary to align with evolving business needs and industry trends.

Score #	Weight	Weighting Criteria
1	**Business Impact (Weight: 25%)**	• Revenue growth attributed to intelligence initiatives • Cost savings achieved through process optimization • Increase in customer satisfaction or retention rates • Improvement in operational efficiency and productivity
2	**Data Management (Weight: 20%)**	• Data quality and integrity measures • Data governance and compliance adherence • Data accessibility and availability • Data security and privacy controls
3	**Analytics and Insights (Weight: 20%)**	• Adoption and utilization of analytics tools and platforms • Accuracy and relevance of insights generated • Actionability of recommendations provided • Integration of analytics into decision-making processes
4	**Innovation and Adaptability (Weight: 15%)**	• Introduction of new data-driven products/services • Implementation of emerging technologies (AI, ML, IoT, etc.) • Ability to adapt to changing market conditions • Proactive identification of new opportunities or risks
5	**Organizational Alignment (Weight: 10%)**	• Integration of intelligence initiatives with overall business strategy • Support and commitment from executive leadership • Collaboration and knowledge-sharing across departments • Alignment of goals and incentives with intelligence initiatives

Score #	Weight	Weighting Criteria
6	**Talent and Skills (Weight: 10%)**	• Availability of skilled data scientists and analysts • Training and development programs for employees • Recruitment and retention of top data and analytics talent • Cross-functional collaboration and knowledge transfer
7	**Measurement and Evaluation (Weight: 10%)**	• Defined key performance indicators (KPIs) for intelligence initiatives • Regular monitoring and reporting of KPIs • Evaluation of performance against set targets and benchmarks • Continuous improvement based on feedback and insights

Key Performance Indicators (KPIs) for Decision Intelligence Generative AI

Key Performance Indicators (KPIs) for decision intelligence generative AI systems can help organizations evaluate the effectiveness and performance of AI-powered decision-making models. These indicators provide insights into the accuracy, efficiency, and impact of decision-making processes facilitated by generative AI systems. Here are some examples of KPIs for decision intelligence generative AI:

Indicator	Metric	Measurement
1	**Decision Accuracy:**	Measures the accuracy of decisions made by the generative AI system compared to human decision-making or predefined benchmarks. It assesses the system's ability to generate high-quality decisions.
2	**Decision Speed:**	Tracks the time taken by the generative AI system to generate decisions. It evaluates the system's efficiency in providing timely decision support, enabling real-time or near-real-time decision-making.
3	**Decision Consistency:**	Evaluates the level of consistency and coherence in the decisions generated by the AI system across different scenarios or iterations. It helps ensure that the system produces reliable and consistent outcomes.
4	**Decision Confidence Level:**	Assesses the level of confidence or certainty associated with the decisions made by the generative AI system. It provides insights into the system's reliability and helps identify areas where human intervention may be necessary.
5	**Decision Bias and Fairness:**	Measures the fairness and impartiality of decisions made by the generative AI system. It helps detect and mitigate any biases or discriminatory patterns in decision-making, promoting ethical and unbiased outcomes.
6	**Decision Explanation and Interpretability:**	Evaluates the system's ability to explain and provide transparent reasoning behind the decisions it generates. It enables users to understand and trust the AI-generated decisions, fostering better adoption and acceptance.

Indicator	Metric	Measurement
7	**Decision Impact and Value:**	Assesses the impact and value of the decisions made by the generative AI system in achieving desired outcomes. It measures factors such as cost savings, revenue generation, risk mitigation, or operational efficiency improvements resulting from the decisions.
8	**User Satisfaction:**	Measures user satisfaction with the generative AI system's decision-making capabilities. It considers factors such as user feedback, surveys, and ratings to assess the system's usability, effectiveness, and user experience.
9	**Decision Adaptability:**	Evaluates the system's ability to adapt and adjust decisions based on changing input data or contextual factors. It measures the agility and flexibility of the generative AI system in responding to dynamic decision-making requirements.
10	**Learning and Continuous Improvement:**	Measures the system's capability to learn from feedback and data, improving decision-making accuracy and performance over time. It assesses the system's ability to leverage data-driven insights for ongoing optimization.

These KPIs help organizations assess the performance, reliability, and impact of decision intelligence generative AI systems. It is important to customize and adapt these KPIs to the specific context, objectives, and industry requirements to ensure meaningful evaluation and optimization of the AI systems.

Key Performance Indicators (KPIs) for Enterprise Strategy Intelligence

Key Performance Indicators (KPIs) for enterprise strategy intelligence are essential metrics that organizations use to measure and assess the effectiveness of their strategic initiatives. These indicators provide insights into the performance and impact of the enterprise's overall strategy. Here are some examples of KPIs for enterprise strategy intelligence:

Indicator	Metric	Measurement
1	Strategic Alignment Index (SAI):	Measures the extent to which the organization's initiatives and activities align with its strategic goals and objectives.
2	Strategic Initiative Success Rate:	Tracks the percentage of strategic initiatives that are successfully implemented and achieve their intended outcomes.
3	Return on Strategy Investment (ROSI):	Calculates the financial return generated from the organization's strategic investments, taking into account factors such as revenue growth, cost savings, and increased market share.
4	Competitive Intelligence Effectiveness:	Assesses the organization's ability to gather, analyze, and leverage intelligence on competitors to make informed strategic decisions and gain a competitive advantage.
5	Market Share Growth:	Measures the percentage increase in the organization's market share over a specific period, indicating the effectiveness of its strategic initiatives in capturing a larger portion of the market.
6	Customer Satisfaction Index (CSI):	Gauges the satisfaction levels of customers or clients with the organization's products, services, and overall customer experience, indicating the impact of strategic decisions on customer relationships.
7	Employee Engagement and Alignment:	Measures employee satisfaction, commitment, and alignment with the organization's strategic goals, as engaged employees are more likely to contribute effectively to strategic initiatives.
8	Innovation Index:	Tracks the number and impact of innovative ideas generated and implemented within the organization, reflecting its ability to foster a culture of innovation and drive strategic growth.
9	Strategic Risk Management Effectiveness:	Evaluates the organization's ability to identify, assess, and mitigate strategic risks, ensuring that the strategy is resilient and adaptive to changing market conditions.

Indicator	Metric	Measurement
10	**Time-to-Market:**	Measures the time it takes for the organization to develop and launch new products or services, indicating its agility and ability to capitalize on market opportunities.

These KPIs provide a holistic view of an organization's strategic performance and help drive informed decision-making, enabling continuous improvement and adaptation to changing business landscapes. It is important to tailor the selection of KPIs to the specific goals and context of the enterprise.

Key Performance Indicators (KPIs) for Assessing Enterprise Initiatives

Key Performance Indicators (KPIs) for assessing enterprise initiatives are metrics that organizations use to measure and evaluate the performance and impact of their initiatives. These indicators provide valuable insights into the effectiveness and success of the initiatives, allowing organizations to make data-driven decisions and improvements. Here are some examples of KPIs for assessing enterprise initiatives:

Indicator	Metric	Measurement
1	Initiative Success Rate:	Measures the percentage of initiatives that achieve their intended goals and objectives. It provides an overall measure of the success of the initiatives undertaken by the organization.
2	Initiative ROI (Return on Investment):	Calculates the financial return generated by an initiative compared to the investment made. It helps assess the financial viability and profitability of the initiative.
3	Time-to-Value:	Tracks the time it takes for an initiative to start delivering measurable value to the organization. It indicates the efficiency of the initiative's implementation and the speed at which it generates results.
4	Benefits Realization:	Assesses the extent to which the anticipated benefits and outcomes of an initiative are realized. It measures the actual value or impact delivered by the initiative compared to the expected benefits.
5	Stakeholder Satisfaction:	Measures the satisfaction levels of key stakeholders, including customers, employees, partners, and shareholders, with the outcomes and impacts of the initiative. It reflects the initiative's ability to meet stakeholder expectations.
6	Adoption and Utilization Rate:	Tracks the rate at which the intended users or beneficiaries adopt and actively utilize the outputs or solutions delivered by the initiative. It indicates the level of acceptance and integration of the initiative within the organization.
7	Process Efficiency:	Measures the efficiency and effectiveness of the processes impacted by the initiative. It evaluates factors such as cycle time reduction, cost savings, error rates, and productivity improvements achieved through the initiative.

Indicator	Metric	Measurement
8	**Risk Mitigation:**	Assesses the initiative's effectiveness in identifying, managing, and mitigating risks. It measures the organization's ability to proactively address risks and minimize their potential impact on the initiative.
9	**Lessons Learned and Knowledge Capture:**	Evaluates the initiative's ability to capture and document lessons learned, best practices, and knowledge gained during its implementation. It helps facilitate organizational learning and continuous improvement.
10	**Strategic Alignment and Contribution:**	Measures the alignment of the initiative with the organization's strategic objectives and its overall contribution to the strategic goals. It assesses how well the initiative supports the organization's long-term vision and direction.

These KPIs provide a comprehensive evaluation of the performance, value, and impact of enterprise initiatives. By monitoring and analyzing these metrics, organizations can identify areas of improvement, make informed decisions, and optimize their initiatives for better outcomes. It is important to customize and align these KPIs to the specific goals, objectives, and nature of the initiatives undertaken by the organization.

Key Performance Indicators (KPIs) For Scoring Enterprise Initiatives

Key Performance Indicators (KPIs) for scoring enterprise initiatives are metrics used to evaluate and score the performance and effectiveness of various initiatives undertaken by an organization. These indicators help assign a numerical value or score to each initiative, enabling organizations to compare and prioritize them based on their performance. Here are some examples of KPIs for scoring enterprise initiatives:

Indicator	Metric	Measurement
1	Financial Impact:	Measures the financial benefits or return on investment (ROI) generated by an initiative. It evaluates factors such as revenue growth, cost savings, profitability, and payback period.
2	Strategic Alignment:	Assesses the extent to which an initiative aligns with the organization's strategic objectives and goals. It evaluates the initiative's contribution to the overall strategic direction and vision.
3	Customer Satisfaction:	Measures the level of customer satisfaction and feedback resulting from the initiative. It evaluates factors such as customer surveys, Net Promoter Score (NPS), customer retention rates, and customer complaints or compliments.
4	Operational Efficiency	Assesses the efficiency and effectiveness of the initiative in optimizing operational processes and workflows. It evaluates metrics such as cycle time reduction, resource utilization, productivity improvement, and waste reduction.
5	Innovation and Creativity:	Measures the level of innovation and creativity exhibited by the initiative. It evaluates factors such as the introduction of new products or services, adoption of new technologies or processes, and the number of patents or intellectual property generated.
6	Risk Management:	Evaluates the effectiveness of the initiative in identifying, assessing, and mitigating risks. It considers factors such as risk identification, risk mitigation strategies implemented, and the impact of risk events on the initiative's progress.
7	Stakeholder Engagement:	Assesses the level of engagement and satisfaction of key stakeholders, including employees, customers, partners, and shareholders. It considers factors such as stakeholder surveys, feedback mechanisms, and stakeholder involvement in decision-making.

Indicator	Metric	Measurement
8	**Timeliness and Execution:**	Measures the effectiveness of change management efforts associated with the initiative. It evaluates factors such as employee adoption and acceptance of the initiative, training programs implemented, and communication strategies employed.
9	**Change Management:**	Evaluates the initiative's ability to capture and document lessons learned, best practices, and knowledge gained during its implementation. It helps facilitate organizational learning and continuous improvement.
10	**Lessons Learned and Continuous Improvement:**	Assesses the initiative's ability to learn from past experiences and apply continuous improvement practices. It considers factors such as post-implementation reviews, corrective actions taken, and feedback loops for iterative enhancements.

By scoring initiatives based on these KPIs, organizations can objectively evaluate and compare their performance, identify areas for improvement, and make informed decisions regarding resource allocation and prioritization. The specific KPIs and their weightage may vary depending on the organization's industry, strategic objectives, and specific initiative requirements.

Key Performance Indicators (KPIs) For Prioritizing Enterprise Initiatives

Key Performance Indicators (KPIs) for prioritizing enterprise initiatives are essential metrics that organizations use to evaluate and prioritize their initiatives based on their strategic importance and potential impact. These indicators help organizations make informed decisions about allocating resources and focus on the initiatives that will deliver the most value. Here are some examples of KPIs for prioritizing enterprise initiatives:

Indicator	Metric	Measurement
1	**Strategic Alignment Score:**	Measures the extent to which an initiative aligns with the organization's strategic objectives and priorities. It helps assess whether the initiative supports the overall strategic direction of the enterprise.
2	**Business Value Potential:**	Evaluates the estimated financial or strategic value that an initiative can generate for the organization. It considers factors such as revenue growth, cost savings, market expansion, competitive advantage, and customer satisfaction.
3	**Feasibility and Resource Requirements:**	Assesses the feasibility of executing an initiative and the resources (financial, human, technological) required for its implementation. It helps determine if the organization has the necessary capabilities and capacity to undertake the initiative.
4	**Impact on Key Performance Areas:**	Measures the potential impact of an initiative on key performance areas or critical success factors of the organization. These areas may include revenue growth, customer acquisition and retention, operational efficiency, employee productivity, and innovation.
5	**Risk versus Reward Analysis:**	Evaluates the risks associated with an initiative against the potential rewards it offers. It helps assess the level of risk tolerance within the organization and prioritize initiatives that strike a balance between risk mitigation and value creation.
6	**Market Demand and Opportunity:**	Assesses the market demand and growth potential for the products, services, or solutions that an initiative aims to deliver. It considers factors such as market size, customer needs, competitive landscape, and emerging trends.

Indicator	Metric	Measurement
7	**Strategic Fit and Alignment with Core Competencies:**	Measures how well an initiative aligns with the organization's core competencies and its ability to leverage existing strengths and capabilities. It helps identify initiatives that capitalize on the organization's competitive advantages.
8	**Time Sensitivity:**	Evaluates the urgency or time sensitivity of an initiative. It considers factors such as market timing, competitive pressures, regulatory requirements, and internal deadlines. Initiatives with time-sensitive opportunities or constraints may receive higher priority
9	**Stakeholder Influence and Support:**	Assesses the level of support and influence that key stakeholders, such as executives, customers, partners, and employees, have on an initiative. It helps prioritize initiatives that have strong stakeholder buy-in and alignment.
10	**Alignment with Industry or Market Trends:**	Considers the relevance and alignment of an initiative with current industry or market trends. It helps organizations stay ahead of the curve and prioritize initiatives that address emerging challenges or capitalize on market opportunities.

These KPIs provide a structured framework for evaluating and prioritizing enterprise initiatives, ensuring that resources are allocated to initiatives that are strategically aligned and have the potential to deliver significant value to the organization. The selection and weightage of these KPIs may vary based on the organization's industry, goals, and specific context.

Key Performance Indicators (KPIs) for Optimizing Enterprise Initiatives

Key Performance Indicators (KPIs) for optimizing enterprise initiatives are crucial metrics that help organizations monitor and measure the effectiveness and efficiency of their initiatives. These indicators enable businesses to identify areas of improvement, track progress, and make data-driven decisions to optimize their initiatives. Here are some examples of KPIs for optimizing enterprise initiatives:

Indicator	Metric	Measurement
1	Initiative ROI (Return on Investment):	Measures the financial return generated by an initiative, taking into account the costs incurred and the benefits gained. It helps assess the profitability and value delivered by the initiative.
2	Initiative Cycle Time:	Tracks the time it takes to plan, execute, and complete an initiative, providing insights into the efficiency of the initiative's implementation process.
3	Initiative Cost Variance:	Compares the actual costs of an initiative with the budgeted costs, helping identify cost overruns or savings. It enables organizations to optimize resource allocation and control expenses.
4	Stakeholder Satisfaction:	Assesses the satisfaction levels of key stakeholders, such as customers, employees, and partners, with the outcomes and impacts of an initiative. It reflects the initiative's effectiveness in meeting their expectations and delivering value.
5	Initiative Quality Index:	Measures the quality of the deliverables or outcomes of an initiative, considering factors such as accuracy, completeness, functionality, and compliance. It helps ensure that the initiative meets the required standards and specifications.
6	Initiative Adoption Rate:	Tracks the rate at which the organization's target audience or users adopt and embrace the initiative's outputs or solutions. It indicates the initiative's acceptance and impact on the intended beneficiaries.
7	Initiative Scalability:	Assesses the ability of the initiative to scale and expand its impact across different business units, departments, or geographies. It indicates the potential for wider implementation and optimization of the initiative.

Indicator	Metric	Measurement
8	**Initiative Risk Exposure:**	Evaluates the level of risks associated with an initiative, including factors such as market risks, operational risks, legal and regulatory risks, and reputational risks. It helps identify and mitigate potential threats that may affect the initiative's success.
9	**Initiative Feedback Loop:**	Tracks the feedback and suggestions received from stakeholders regarding the initiative. It helps identify areas for improvement, optimize processes, and enhance the initiative's outcomes.
10	**Initiative Learning and Knowledge Transfer:**	Measures the extent to which knowledge, insights, and lessons learned from the initiative are captured, documented, and shared across the organization. It ensures that valuable knowledge is leveraged to optimize future initiatives.

These KPIs enable organizations to assess the performance, efficiency, and impact of their initiatives, identify bottlenecks, and drive continuous improvement. It is important to align the selection of KPIs with the specific goals, objectives, and nature of the enterprise initiatives to ensure meaningful measurement and optimization.

Decision Intelligence Generative AI Business Case Validation Calculators

Introduction:
The Decision Intelligence Generative AI Business Case Validation Calculator is a tool designed to assist organizations in evaluating the viability and potential return on investment (ROI) of implementing decision intelligence generative AI solutions. Decision intelligence refers to the practice of combining data, analytics, and AI to optimize decision-making processes. This calculator will help businesses determine whether investing in decision intelligence generative AI is financially justified and aligns with their strategic goals.

Key Inputs:
1. Business Objectives: Define the specific goals and objectives that the decision intelligence generative AI solution aims to achieve. This could include improving operational efficiency, enhancing customer experience, reducing costs, or increasing revenue.

2. Data Availability: Assess the availability and quality of relevant data required to train the AI model. Consider factors such as data completeness, accuracy, and security.

3. Implementation Costs: Determine the estimated costs associated with implementing the decision intelligence generative AI solution. This includes expenses related to data preparation, infrastructure, software licenses, AI model development, and integration with existing systems.

4. Timeframe: Define the desired timeframe for implementing the solution and achieving the expected outcomes.

5. Performance Metrics: Identify the key performance indicators (KPIs) that will be used to measure the success of the decision intelligence generative AI solution. These metrics could be related to cost savings, revenue growth, customer satisfaction, or process efficiency.

Calculation Steps:
1. Quantify Benefits: Estimate the potential benefits that the decision intelligence generative AI solution can deliver based on the defined business objectives. This may involve conducting market research, benchmarking against industry standards, or consulting subject matter experts.

2. Assess Costs: Evaluate the costs associated with implementing and maintaining the solution over the defined timeframe. Consider both initial implementation costs and ongoing expenses, such as maintenance, updates, and training.

3. ROI Calculation: Calculate the return on investment (ROI) by comparing the estimated benefits against the total costs. The formula for ROI is:

ROI = (Net Benefits / Total Costs) x 100

4. Sensitivity Analysis: Perform a sensitivity analysis to evaluate the impact of varying key inputs, such as data availability, implementation costs, or performance metrics. This will provide insights into the potential range of outcomes and help assess the robustness of the business case.

Output:
The Decision Intelligence Generative AI Business Case Validation Calculator will provide the following outputs:

1. ROI Analysis: The calculated ROI percentage based on the estimated benefits and costs.

2. Break-even Point: Determine the timeframe required to recover the initial investment and reach the break-even point.

3. Sensitivity Analysis: Insights into the impact of varying key inputs on the ROI and potential outcomes.

4. Decision Recommendation: Provide a recommendation on whether to proceed with the implementation of the decision intelligence generative AI solution based on the calculated ROI and other factors.

Conclusion:
The Decision Intelligence Generative AI Business Case Validation Calculator assists organizations in evaluating the financial viability of implementing decision intelligence generative AI solutions. By considering inputs such as business objectives, data availability, implementation costs, timeframe, and performance metrics, the calculator provides a comprehensive analysis of the potential ROI. This tool helps businesses make informed decisions and prioritize investments that align with their strategic goals, maximizing the benefits of decision intelligence generative AI.

Initiative Prioritization Calculator

The purpose of this calculator is to assist in prioritizing different initiatives based on certain criteria. Here's a simple step-by-step guide to creating the calculator:

Step 1: Define the Criteria:
Determine the criteria you want to use for prioritizing initiatives. These criteria can be based on factors such as strategic alignment, potential impact, feasibility, resources required, and urgency.

Step 2: Assign Weightage:
Assign a weightage to each criterion to reflect its importance relative to others. The weightage should be on a scale of 1 to 10, where a higher weight indicates higher importance.

Step 3: Create a Scoring System:
Define a scoring system for each criterion. For example, you can use a scale of 1 to 5 or 1 to 10 to rate each initiative on each criterion. Make sure to document the meaning of each score to maintain consistency.

Step 4: Collect Data:
Collect the necessary data for each initiative. This may include information such as the initiative's description, potential impact, resources required, and urgency level.

Step 5: Calculate the Scores:
For each initiative, calculate the score for each criterion by multiplying the score by the weightage

assigned to that criterion. Sum up the scores for all criteria to obtain a total score for each initiative.

Step 6: Rank the Initiatives:
Rank the initiatives based on their total scores in descending order. The higher the total score, the higher the priority of the initiative.

Step 7: Review and Refine:
Review the results and consider any additional factors that may influence the prioritization. Refine the criteria, weightage, or scoring system if necessary.

Step 8: Utilize the Calculator:
Use the calculator to input data for new initiatives and calculate their scores. Update the rankings accordingly.

Remember that this is a basic framework, and you can customize it further based on your specific needs and requirements. Additionally, you may consider using spreadsheet software like Prioriti AI to automate the calculations and facilitate data management.

Initiative Scoring Calculator

Initiative scoring calculator that you can use to score and rank initiatives based on different criteria:

Step 1: Define the Criteria:
Identify the criteria that you want to use for scoring initiatives. For example, you can consider criteria such as strategic alignment, potential impact, feasibility, resources required, and urgency.

Step 2: Assign Weightage:
Assign a weightage to each criterion to indicate its importance relative to others. The weightage should be a percentage that adds up to 100%. Allocate higher weightage to criteria that are more crucial for your decision-making process.

Step 3: Create a Scoring System:
Establish a scoring system for each criterion. You can use a scale of 1 to 10, where 1 represents the lowest score and 10 represents the highest score. Ensure that the scoring system is consistent across all criteria.

Step 4: Collect Data:
Gather the necessary data for each initiative, including information relevant to each criterion. This could involve gathering details such as the initiative's description, potential impact, resources required, and urgency level.

Step 5: Calculate the Scores:
For each criterion, assign a score to each initiative based on its alignment or performance in that particular criterion. Multiply the score by the weightage assigned to the criterion. Repeat this process for all criteria. Sum up the weighted scores to obtain a total score for each initiative.

Step 6: Rank the Initiatives:
Rank the initiatives based on their total scores in descending order. The higher the total score, the higher the priority of the initiative.

Step 7: Review and Refine:

Review the results and consider any additional factors that may influence the scoring and prioritization. Refine the criteria, weightage, or scoring system as necessary to improve the accuracy of the results.

Step 8: Utilize the Calculator:

Use the calculator to input data for new initiatives and calculate their scores. Update the rankings accordingly.

You can implement this calculator using a spreadsheet software like Prioriti AI. Each initiative can have a row where you input the scores and weightages, and the calculator can automatically calculate the total score and rank the initiatives based on the provided data.

Remember, this is a basic framework, and you can customize it further based on your specific requirements and the complexity of your decision-making process.

Return on Decision Intelligence (RODI) Calculator

Return on Decision Intelligence (RODI) calculator is a tool used to assess the value and effectiveness of decision intelligence initiatives. Decision intelligence refers to the use of data, analytics, and decision-making frameworks to improve the quality and outcomes of business decisions. Here's a simplified guide to creating an RODI calculator:

Step 1: Define Input Variables:

Identify the key input variables that contribute to the RODI calculation. These variables typically include factors such as:

1. Cost of the decision intelligence initiative (e.g., implementation costs, data acquisition costs, technology investments, and ongoing maintenance expenses).

2. Expected improvements or benefits resulting from the decision intelligence initiative (e.g., increased revenue, cost savings, risk reduction, improved efficiency, or better strategic outcomes).

3. Timeframe for realizing the benefits (e.g., annual or monthly basis).

Step 2: Calculate RODI:

Once you have identified the input variables, you can calculate the RODI using the following formula:
$$RODI = ((Benefits - Costs) / Costs) * 100$$

Where:

- RODI is the return on decision intelligence percentage.
- Benefits represent the total value gained from the decision intelligence initiative.
- Costs refer to the total costs associated with implementing and maintaining the initiative.

Step 3: Collect Data:
Gather the necessary data to populate the input variables. This may involve analyzing historical data, conducting surveys, consulting with subject matter experts, or estimating potential benefits based on industry benchmarks.

Step 4: Input Data and Calculate RODI:
Create an interface where you can input the data for each variable. This can be a form or a spreadsheet. Once the data is inputted, use the RODI formula to calculate the RODI based on the provided values.

Step 5: Interpret the Results:
Display the calculated RODI to the user, along with any other relevant information, such as the expected benefits, costs, and timeframe for realizing the benefits. It's also useful to provide a summary or analysis of the results to help stakeholders interpret the findings.

Step 6: Review and Refine:
Review the results and consider any additional factors that may impact the accuracy of the RODI calculation. Refine the input variables, formulas, or data sources if necessary to improve the accuracy and reliability of the calculator.

Step 7: Utilize the Calculator:
Use the calculator to evaluate different decision intelligence initiatives by inputting the relevant data and calculating their respective RODIs. This allows you to compare and prioritize initiatives based on their potential returns.

It's important to note that decision intelligence initiatives can have a wide range of benefits and costs, and quantifying the impact of improved decision-making can sometimes be challenging. This simplified framework provides a starting point, but you can customize and expand the calculator based on your specific needs and requirements.

Determining The Timing For Implementing Enterprise Strategy Intelligence Requires Careful Consideration Of Several Criteria. Here Are Some Key Factors To Assess:

1. **Strategic Objectives:**
   - Are there specific strategic objectives or challenges that can be addressed through intelligence-driven insights?
   - Is there a need to enhance decision-making processes and align them with strategic goals?
   - Are there opportunities to gain a competitive advantage by leveraging data and intelligence?
2. **Data Availability and Quality:**
   - Is the organization collecting and storing relevant data necessary for intelligence initiatives?
   - Is the data of sufficient quality, accuracy, and completeness to derive meaningful insights?
   - Are there data gaps that need to be addressed before implementing enterprise strategy intelligence?

3. Technological Readiness:

- Does the organization have the necessary technology infrastructure to support intelligence initiatives?
- Are there appropriate analytics tools and platforms available for data analysis and visualization?
- Can existing systems integrate and share data effectively for intelligence purposes?

4. Resource Allocation:

- Is the organization willing and able to allocate sufficient resources (financial, human, and technological) to implement and sustain enterprise strategy intelligence?
- Are there budgetary constraints or competing priorities that may impact the implementation timeline?

5. Organizational Culture:

- Is there a culture that embraces data-driven decision-making and values intelligence?
- Are employees receptive to change and willing to adopt new ways of working?
- Is there executive support and commitment to drive the adoption of enterprise strategy intelligence?

6. Risk Assessment:

- Are there any risks associated with implementing intelligence initiatives, such as data security or privacy concerns?
- Can the organization manage and mitigate these risks effectively?
- Are there regulatory or compliance requirements that need to be considered?

7. Stakeholder Engagement:

- Are key stakeholders, including senior leadership, onboard with the idea of enterprise strategy intelligence?
- Do stakeholders understand the potential benefits and are they supportive of the initiative?
- Is there active participation and engagement from relevant departments and teams?

8. Industry and Market Dynamics:

- Are there industry trends or market conditions that make it necessary to adopt enterprise strategy intelligence?
- Are competitors or peers already leveraging intelligence, and is there a risk of falling behind?

By evaluating these criteria, an organization can assess its readiness to implement enterprise strategy intelligence. It is important to conduct a thorough analysis and consider the potential benefits, challenges, and resources required before making a decision on the timing of implementation.

Creating your Enterprise Strategy Intelligence Strategic Fit and Alignment with Core Competencies

Strategic Fit and Alignment with Core Competencies is a critical aspect of enterprise strategy intelligence. It involves ensuring that the strategic direction of an organization aligns with its core competencies, capabilities, and resources to achieve sustainable competitive advantage. Here is a framework to help you create an effective strategy that leverages your core competencies:

1. Understand Your Core Competencies:
- Identify and define your organization's core competencies, which are unique strengths and capabilities that set you apart from competitors.
- Evaluate the resources, skills, technologies, and knowledge that contribute to your core competencies.
- Determine how your core competencies can create value for your customers and support your strategic goals.

2. Define Strategic Direction:
- Clarify your organization's mission, vision, and long-term strategic goals.
- Conduct a comprehensive analysis of the external environment, including market trends, customer needs, and competitor strategies.
- Identify opportunities and threats that align with your core competencies.

3. Conduct a Gap Analysis:
- Assess the current state of your organization's capabilities and resources.
- Identify the gaps between your current position and the desired strategic outcomes.
- Determine which core competencies need to be enhanced, developed, or acquired to bridge the gaps.

4. Prioritize Strategic Initiatives:
- Evaluate various strategic initiatives based on their alignment with your core competencies and potential for creating competitive advantage.
- Prioritize the initiatives that best leverage and capitalize on your core competencies.
- Consider factors such as market demand, feasibility, resource availability, and potential risks.

5. Develop an Implementation Plan:
- Define clear objectives, action plans, timelines, and performance indicators for each strategic initiative.
- Allocate resources, both financial and human, to support the implementation of the chosen initiatives.
- Establish a monitoring and evaluation process to track progress, make necessary adjustments, and ensure alignment with core competencies.

6. Foster Organizational Alignment:
- Communicate the strategic direction and the rationale behind it to all levels of the organization.
- Engage employees in the strategy development process and provide opportunities for feedback and input.

- Align organizational structures, processes, and incentives to support the execution of the strategic initiatives.

7. Continuously Monitor and Adapt:
- Regularly assess the performance of strategic initiatives and their alignment with core competencies.
- Monitor changes in the external environment and make adjustments to the strategy as needed.
- Foster a culture of learning, innovation, and agility to stay ahead of competitors.

By following this framework, you can ensure that your enterprise strategy aligns closely with your core competencies, enabling your organization to achieve sustainable competitive advantage and long-term success.

Business Case for Implementing Enterprise Strategy Intelligence

By presenting a comprehensive business case that demonstrates the need, benefits, implementation plan, costs, and stakeholder alignment, the organization can justify the implementation of enterprise strategy intelligence and gain support from key decision-makers.

Case	Action
Executive Summary:	• Describe the key points of the business case, including the need for enterprise strategy intelligence, its potential benefits, and the expected return on investment.
Business Need:	• Identify the strategic challenges or opportunities that can be addressed through intelligence-driven insights. • Highlight the limitations of current decision-making processes and the potential risks of not leveraging intelligence.
Objectives:	• Clearly define the objectives of implementing enterprise strategy intelligence, such as improving strategic decision-making, enhancing competitiveness, or driving revenue growth. • Align the objectives with the overall business strategy and long-term goals.
Benefits:	• Quantify the potential benefits of enterprise strategy intelligence, such as improved operational efficiency, cost savings, revenue growth, and enhanced customer satisfaction. • Highlight specific examples or case studies showcasing the benefits realized by other organizations.
Key Features and Capabilities:	• Describe the key features and capabilities of enterprise strategy intelligence, including data integration, advanced analytics, predictive modeling, and visualization tools. • Explain how these features can enable better insights, actionable recommendations, and informed decision-making.
Data and Technology Requirements:	• Assess the organization's current data infrastructure and identify any gaps or challenges. • Outline the data requirements and the technology needed to support enterprise strategy intelligence initiatives. • Discuss potential data sources, data quality measures, and necessary technology investments.
Implementation Plan:	• Provide a high-level implementation plan, including timelines, milestones, and key activities. • Identify any dependencies or potential risks associated with the implementation. • Allocate resources and specify the team or department responsible for executing the plan.

Case	Action
Cost and Investment:	• Estimate the costs associated with implementing enterprise strategy intelligence, including technology acquisition, infrastructure upgrades, training, and staffing. • Conduct a cost-benefit analysis to demonstrate the return on investment (ROI) and the payback period. • Highlight potential cost savings or revenue growth that can be attributed to intelligence-driven decision-making.
Risk Assessment:	• Identify potential risks and challenges associated with implementing enterprise strategy intelligence, such as data security, privacy concerns, or resistance to change. • Develop a risk mitigation plan and outline strategies to address each identified risk
Stakeholder Alignment:	• Identify the key stakeholders and their roles in the implementation of enterprise strategy intelligence. • Discuss the benefits and value proposition for each stakeholder group, such as improved insights for executives, enhanced performance for operational teams, and increased customer satisfaction. • Highlight any endorsements or support received from key stakeholders.
Conclusion:	• Summarize the business case, emphasizing the strategic importance, potential benefits, and ROI of implementing enterprise strategy intelligence. • Reinforce how enterprise strategy intelligence aligns with the organization's goals and competitive advantage.

Implementation Plan for Enterprise Strategy Intelligence with Generative AI

Task	Description
Define Objectives and Scope:	• Clearly define the objectives of implementing enterprise strategy intelligence with generative AI, such as generating innovative strategies, optimizing resource allocation, or predicting market trends. • Determine the scope of the implementation, including the specific areas or departments where generative AI will be applied.
Data Assessment and Preparation:	• Identify the data required for generative AI models, such as historical market data, financial data, customer data, and industry reports. • Assess the availability, quality, and completeness of the required data. • Prepare the data by cleansing, integrating, and organizing it in a format suitable for generative AI algorithms.
Technology Infrastructure:	• Evaluate the existing technology infrastructure and determine if additional infrastructure is needed to support generative AI. • Identify and acquire the necessary generative AI tools and platforms. • Set up the required hardware and software infrastructure to enable generative AI processing.
Data Modeling and Training:	• Select appropriate generative AI algorithms and models based on the objectives and data characteristics. • Train the generative AI models using the prepared data. • Fine-tune and optimize the models to ensure accurate and reliable outputs.
Integration and Deployment:	• Integrate the generative AI models into the existing enterprise strategy intelligence framework or decision-making processes. • Develop APIs or interfaces to enable seamless integration with other systems or applications. • Test the integration and deployment to ensure proper functioning and alignment with business requirements.
User Training and Adoption:	• Provide training and education to the users who will be utilizing the generative AI outputs. • Demonstrate how to interpret and leverage generative AI-generated insights for strategic decision-making. • Foster a culture of data-driven decision-making and encourage user adoption of generative AI capabilities.

Task	Description
Performance Monitoring and Optimization:	• Implement mechanisms to monitor the performance of generative AI models and the impact on strategic decision-making. • Continuously assess the accuracy, reliability, and effectiveness of generative AI outputs. • Incorporate feedback and insights from users to optimize and refine the generative AI algorithms and models.
Governance and Ethics:	• Establish governance mechanisms to ensure ethical and responsible use of generative AI. • oevelop guidelines and policies for data privacy, security, and compliance. • Regularly review and update the governance framework to align with evolving regulations and industry standards.
Continuous Improvement and Innovation:	• Foster a culture of continuous improvement by encouraging feedback, learning, and experimentation. • Explore opportunities for further innovation and expansion of generative AI applications in enterprise strategy intelligence. • Stay updated with advancements in generative AI technologies and methodologies to leverage the latest capabilities.
Evaluation and Measurement:	• Define key performance indicators (KPIs) to evaluate the effectiveness and impact of generative AI on strategic decision-making. • Monitor and measure the performance of generative AI initiatives against defined KPIs. • Conduct periodic assessments and audits to assess the ROI and identify areas for improvement.
Collaboration and Knowledge Sharing:	• Encourage collaboration and knowledge sharing across teams and departments involved in generative AI and enterprise strategy intelligence. • Establish forums, workshops, or communities of practice to facilitate the exchange of best practices and lessons learned. • Foster a culture of collaboration and cross-functional learning to drive innovation and maximize the value of generative AI.

The implementation plan should be tailored to the organization's specific requirements, resources, and timeline. Regular monitoring, evaluation, and adjustment of the plan will ensure the successful integration of generative AI into enterprise strategy intelligence initiatives.

Risk Assessment for Implementing Enterprise Strategy Intelligence with Generative AI (Risk and Mitigation)

Item	Description
Data Quality and Reliability:	• Risk: Inaccurate or incomplete data inputs may lead to flawed generative AI outputs and unreliable strategic insights. • Mitigation: Implement rigorous data quality control measures, including data cleansing, validation, and verification. Regularly monitor and maintain data integrity throughout the process.
Algorithmic Bias:	• Risk: Generative AI algorithms may inadvertently introduce biases based on the data inputs, leading to biased strategic recommendations or decisions. • Mitigation: Conduct thorough testing and validation of generative AI models to identify and address potential biases. Continuously monitor and update algorithms to minimize bias and promote fairness.
Ethical and Legal Considerations:	• Risk: The use of generative AI may raise ethical concerns related to privacy, security, intellectual property, and compliance with regulations. • Mitigation: Develop robust governance frameworks and policies to ensure ethical and responsible use of generative AI. Comply with relevant privacy and data protection regulations. Conduct legal reviews to address intellectual property and copyright issues.
Interpretability and Explainability:	• Risk: Generative AI models may produce complex or abstract outputs that are difficult to interpret or explain, making it challenging for stakeholders to trust and act upon the generated insights. • Mitigation: Implement techniques for interpretability and explainability of generative AI outputs, such as model visualization, feature importance analysis, or generating explanations alongside recommendations.
Security and Data Privacy:	• Risk: Generative AI implementation may introduce vulnerabilities or expose sensitive data to security breaches, potentially compromising the confidentiality or integrity of strategic information. • Mitigation: Implement robust security measures, including access controls, encryption, secure data storage, and regular security audits. Adhere to data privacy regulations and anonymize sensitive data when possible.

Item	Description
Adoption and Change Management:	• Risk: Resistance to change or lack of user adoption may hinder the successful implementation and utilization of generative AI in strategic decision-making processes. • Mitigation: Develop a comprehensive change management plan, including clear communication, training programs, and user engagement initiatives. Involve key stakeholders early on and address their concerns to promote buy-in and adoption.
Performance and Scalability:	• Risk: Generative AI models may face performance limitations or scalability challenges when dealing with large datasets or complex strategic scenarios, impacting the efficiency and effectiveness of decision-making processes. • Mitigation: Conduct thorough performance testing and capacity planning to ensure the generative AI system can handle the anticipated data volume and complexity. Continuously monitor and optimize performance as the system scales.
Vendor Reliability and Dependency:	• Risk: Dependency on external vendors or third-party solutions for generative AI capabilities may introduce risks related to vendor reliability, support, updates, and potential business disruptions. • Mitigation: Conduct due diligence when selecting vendors, evaluate their reputation, financial stability, support services, and roadmap. Have contingency plans in place to address any disruptions or changes in vendor relationships.

Regular risk assessments and ongoing monitoring are crucial to identify emerging risks, adapt mitigation strategies, and ensure the secure and successful implementation of enterprise strategy intelligence with generative AI.

SWOT Analysis for Enterprise Strategy Intelligence

It's important to note that the specific strengths, weaknesses, opportunities, and threats may vary depending on the organization, industry, and market conditions. Conducting a thorough analysis tailored to the specific context is recommended.

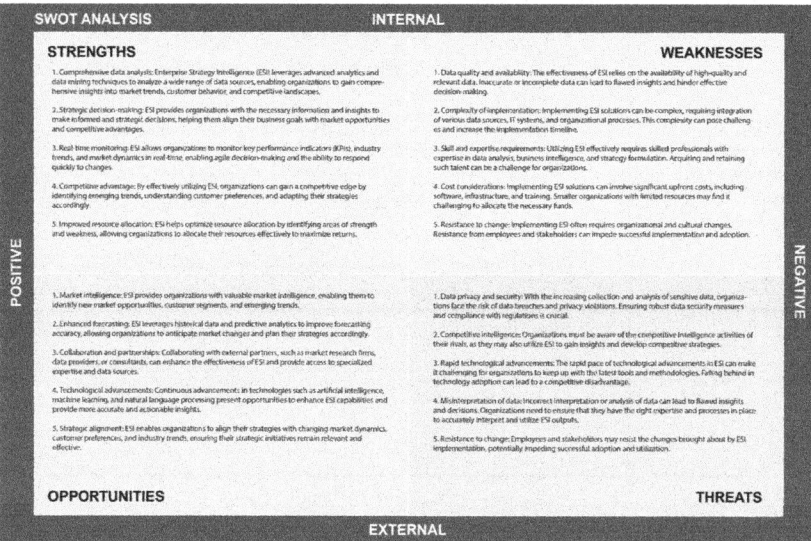

SWOT Analysis for Optimizing Enterprise Initiatives

It's important to note that the specific strengths, weaknesses, opportunities, and threats may vary depending on the organization, industry, and market conditions. Conducting a thorough analysis tailored to the specific context is recommended.

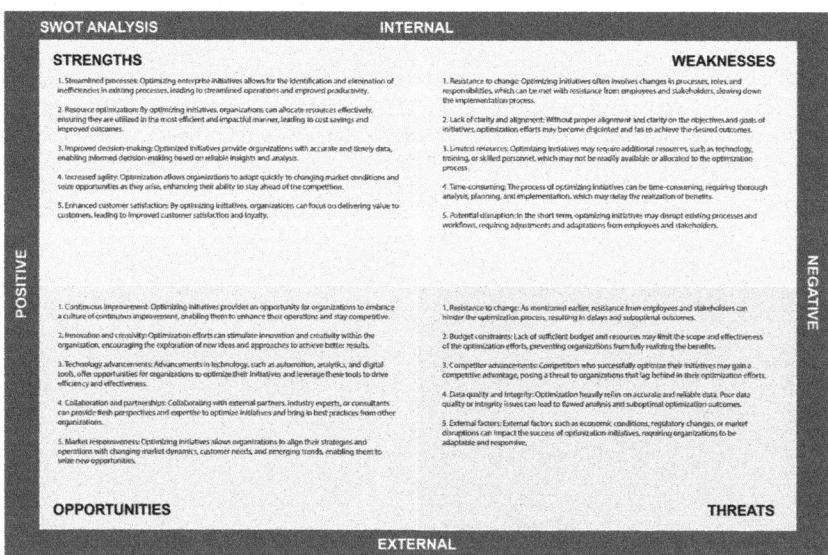

"The consultants have delivered their final report."

Figure 19

Chapter 18: Final

Closing Reflections: The Convergence of Generative AI and Prioritization Decision Intelligence

As we close this exploration of generative AI and prioritization decision intelligence, it's essential to reflect on the journey we've undertaken and the transformative potential these technologies hold. The rapid advancements in artificial intelligence have not only changed the technological landscape but also fundamentally reshaped how we think about creativity, decision-making, and the very nature of intelligence.

The Promise and Perils of Generative AI

Generative AI, with its ability to create text, images, music, and even entire virtual environments, has ushered in a new era of creativity and innovation. It has opened doors to possibilities that were once the domain of science fiction. Artists are collaborating with algorithms to produce novel art forms, writers are using AI to craft narratives that push the boundaries of traditional storytelling, and businesses are leveraging generative models to design products and services that are more personalized and efficient.

However, with great power comes great responsibility. The potential for misuse of generative AI is significant, ranging from deepfakes and misinformation to ethical concerns about authorship and originality. As we harness the power of generative AI, it is imperative that we establish robust frameworks for ethical use, ensuring that these technologies enhance human creativity rather than undermine it.

The Crucial Role of Prioritization Decision Intelligence

In a world increasingly driven by data and automation, the ability to prioritize effectively has never been more critical. Prioritization decision intelligence involves making strategic decisions about where to focus attention and resources, ensuring that the most important tasks and objectives are addressed. This skill is crucial for individuals, organizations, and societies as a whole.

The integration of prioritization decision intelligence with AI can lead to more effective and efficient decision-making processes. By leveraging AI's analytical capabilities, we can identify patterns and insights that inform better prioritization. This symbiosis can help us navigate the complexities of modern life, balancing short-term demands with long-term goals, and optimizing outcomes in an increasingly dynamic environment.The Synergy of Generative AI and Prioritization Intelligence

The intersection of generative AI and prioritization decision intelligence represents a powerful synergy. Generative AI can produce a plethora of options, ideas, and solutions, while prioritization intelligence helps us discern which of these possibilities hold the greatest value. Together, they enable a more holistic approach to problem-solving and innovation.

For example, in healthcare, generative AI can develop a range of treatment plans based on a patient's unique genetic profile, while prioritization decision intelligence can help determine which plan is most likely to be effective. In business, generative AI can generate numerous product prototypes, and prioritization decision intelligence can guide decision-makers in selecting the prototypes with the highest market potential.

Looking Forward

As we move forward, the collaboration between human intelligence and artificial intelligence will be pivotal. Embracing the capabilities of generative AI and prioritization decision intelligence will require a mindset that values continuous learning, ethical consideration, and strategic foresight. We must remain vigilant about the potential challenges and strive to create a future where AI serves as a catalyst for human advancement.

In closing, the journey into the realms of generative AI and prioritization decision intelligence is just beginning. The insights and tools we have explored are the foundation upon which we can build a more innovative, efficient, and equitable world. By understanding and harnessing these technologies, we have the opportunity to unlock new potentials and address some of the most pressing challenges of our time.

The future is not a distant horizon but a canvas waiting to be painted with the brushstrokes of human ingenuity and artificial intelligence. Let us proceed with curiosity, caution, and a commitment to leveraging these powerful tools for the greater good.

Glossary of terms

Glossary of Terms common associated with and with for Generative AI

1. Generative AI: A branch of artificial intelligence that focuses on creating new content, models, or solutions. Generative AI algorithms can generate data or content that resembles real examples based on patterns and information learned during training.

2. Generative Adversarial Networks (GANs): A type of generative AI model consisting of two neural networks: a generator network and a discriminator network. The generator network creates synthetic data samples, while the discriminator network learns to differentiate between real and synthetic data.

3. Variational Autoencoders (VAEs): A type of generative AI model that learns the underlying distribution of data. VAEs aim to encode the input data into a low-dimensional latent space and then decode it back to generate similar but not identical samples.

4. Deep Learning: A subfield of machine learning that focuses on training neural networks with multiple layers to learn and extract complex patterns and representations from data. Deep learning models are widely used in generative AI to capture intricate relationships and generate high-dimensional content.

5. Latent Space: A low-dimensional representation of data that captures the underlying structure and features. In generative AI, latent space refers to the space in which generative models operate to generate new data or content.

6. Training Data: The dataset used to train generative AI models. It consists of real-world examples that the model learns from to generate synthetic data or content.

7. Synthetic Data: Data generated by generative AI models that resembles real-world examples but is not derived directly from real observations. Synthetic data is useful for augmenting existing datasets, simulating scenarios, or expanding the range of available data for analysis.

8. Overfitting: A phenomenon in machine learning where a model becomes too closely fitted to the training data and fails to generalize well to new, unseen data. Overfitting can result in poor performance and limited diversity in the generated output of generative AI models.

9. Mode Collapse: A problem in generative AI where the model collapses to producing a limited set of similar or repetitive outputs, lacking diversity or failing to capture the full range of possibilities.

10. Transfer Learning: A technique in which knowledge learned from one task or domain is transferred and applied to another related task or domain. Transfer learning can be beneficial in generative AI to leverage pre-trained models and accelerate training or improve performance on a specific task.

11. Bias: In generative AI, bias refers to systematic errors or prejudices that may be present in the generated output due to biases in the training data or the model architecture. Addressing and mitigating bias is crucial for fair and unbiased generative AI applications.

12. Fine-tuning: The process of further training a pre-trained generative AI model on a specific task or domain to adapt it to specific requirements or improve its performance on a particular objective.

13. Inference: The process of using a trained generative AI model to generate new data or content based on input or latent variables.

14. Adversarial Attacks: Manipulations or modifications of input data to deceive or mislead generative AI models, leading to unexpected or undesired outputs. Adversarial attacks aim to exploit vulnerabilities in the model's training or architecture.

15. Ethical AI: The practice of ensuring that generative AI models and applications adhere to ethical principles, respect human values, and avoid harmful or discriminatory outcomes. Ethical considerations in generative AI involve fairness, transparency, accountability, and privacy protection.

Remember, this glossary provides an overview of key terms related to generative AI, and further exploration may be needed to fully understand the nuances and intricacies of these concepts.

Glossary Of Terms Commonly Used In The Context Of Enterprise Strategy Intelligence

1. Enterprise Strategy Intelligence: The process of gathering, analyzing, and utilizing data and insights to inform and support strategic decision-making within an organization.

2. Key Performance Indicators (KPIs): Quantifiable metrics used to evaluate the performance and progress of an organization or specific initiatives. KPIs provide objective measures of success and help monitor the effectiveness of strategic actions.

3. Strategic Planning: The process of defining an organization's long-term goals, determining the strategies to achieve those goals, and allocating resources to execute the strategies effectively.

4. Competitive Intelligence: The systematic collection and analysis of information about competitors and the competitive landscape to gain insights and inform strategic decision-making. It helps organizations understand the strengths, weaknesses, opportunities, and threats in their market.

5. SWOT Analysis: A strategic planning technique that assesses an organization's internal strengths and weaknesses, as well as external opportunities and threats. SWOT analysis helps identify key areas for strategic focus and provides a framework for decision-making.

6. Market Research: The process of collecting and analyzing data about target markets, customers, and competitors to gain insights into market trends, customer needs, and competitive landscape. It informs strategic decisions related to market positioning, product development, and customer targeting.

7. Data Analytics: The practice of analyzing raw data to derive meaningful insights and support decision-making. It involves techniques such as statistical analysis, data mining, and predictive modeling to uncover patterns, trends, and relationships within data sets.

8. Data Visualization: The representation of data and insights in a visual format, such as charts, graphs, and infographics. Data visualization helps communicate complex information in a more accessible and understandable way, facilitating data-driven decision-making.

9. Scenario Planning: A strategic planning technique that involves creating and analyzing multiple future scenarios to anticipate potential outcomes and develop strategies to address them. It helps organizations prepare for uncertainties and make more informed decisions in dynamic environments.

10. Performance Management: The ongoing process of setting goals, tracking progress, and evaluating performance against targets. It involves the use of performance metrics, feedback mechanisms, and performance reviews to align activities with strategic objectives and drive continuous improvement.

11. Business Intelligence (BI): The use of technology, tools, and processes to collect, analyze, and present business information and insights. BI helps organizations gain a deeper understanding of their operations, performance, and market dynamics to support strategic decision-making.

12. Predictive Analytics: The practice of using historical data, statistical algorithms, and machine learning techniques to make predictions about future events or outcomes. Predictive analytics enables organizations to anticipate trends, forecast demand, and optimize decision-making.

13. Decision Support Systems (DSS): Computer-based tools and technologies that assist in decision-making by providing relevant information, analysis, and models. DSS integrates data, models, and user interfaces to support strategic decision-making processes.

14. Data Governance: The framework, policies, and processes that ensure the availability, usability, integrity, and security of organizational data. Data governance aims to establish accountability, define data standards, and maintain data quality for effective strategy intelligence initiatives.

15. Executive Dashboard: A visual display of key performance metrics and indicators that provide a real-time snapshot of an organization's performance. Executive dashboards offer a consolidated view of critical data to support strategic decision-making at the executive level.

These are just a few of the terms commonly used in the field of enterprise strategy intelligence. The specific terminology may vary based on industry, organization, and context.

Glossary Of Terms Commonly Used In The Context Of Product Decision Intelligence

1. Product Decision Intelligence: The process of utilizing data, insights, and analytics to inform and optimize product-related decision-making throughout the product lifecycle.

2. Market Research: The process of collecting and analyzing data about target markets, customer needs, and competitive landscape to gain insights that inform product development and strategy.

3. Customer Segmentation: The practice of dividing a customer base into distinct groups or segments based on shared characteristics such as demographics, behaviors, or preferences. Customer segmentation helps tailor product strategies and marketing efforts to specific customer segments.

4. User Personas: Fictional representations of target users or customer segments, based on research and data. User personas capture key characteristics, goals, and needs of different user types, aiding in product design and decision-making.

5. Feature Prioritization: The process of evaluating and ranking product features based on their importance, value, and impact on user experience. Feature prioritization helps allocate resources effectively and deliver the most impactful features first.

6. A/B Testing: A method of comparing two versions (A and B) of a product or feature to determine which performs better in terms of user engagement, conversion rates, or other desired metrics. A/B testing provides empirical data to guide product decision-making.

7. Minimum Viable Product (MVP): The most basic version of a product that includes only essential features and functionalities. An MVP is developed and released to gather user feedback and validate assumptions before investing in additional development.

8. Product-Market Fit: The state where a product meets the needs and expectations of its target market. Product-market fit is achieved when there is a strong alignment between the product's value proposition and the customers' demands.

9. Churn Rate: The rate at which customers stop using or cancel a product or service. Churn rate is an important metric in measuring customer retention and product performance.

10. Customer Lifetime Value (CLTV): The predicted net profit generated by a customer throughout their entire relationship with a company. CLTV helps assess the long-term value of acquiring and retaining customers.

11. Competitive Analysis: The process of evaluating and understanding the strengths, weaknesses, strategies, and market positioning of competitors. Competitive analysis provides insights into the competitive landscape and informs product decision-making.

12. Product Analytics: The practice of collecting and analyzing data related to product usage, user behavior, and performance. Product analytics helps uncover trends, user insights, and opportunities for product improvement.

13. Roadmap: A visual representation of the product's strategic direction and planned features or enhancements over time. Product roadmaps help align stakeholders and communicate the product vision and priorities.

14. Innovation Pipeline: A systematic process of generating and evaluating ideas, concepts, and innovations for potential inclusion in the product portfolio. The innovation pipeline ensures a continuous flow of new ideas to drive product growth and competitiveness.

15. Product-Market Strategy: The overall approach and plan for positioning a product in the market and capturing value. The product-market strategy encompasses target market selection, competitive differentiation, pricing, distribution, and marketing tactics.

These are just a few terms commonly used in the field of product decision intelligence. The specific terminology may vary based on the industry, organization, and product domain.

Endnotes

Who is Prioriti AI? Introducing Prioriti AI: Empowering Global Enterprises with AI-Driven Decision Intelligence.

Are you an executive in a Global 2000 enterprise looking to navigate the complex landscape of initiatives and make informed decisions with confidence? Look no further than Prioriti AI, the cutting-edge Generative AI-driven SaaS platform designed specifically to assist you in assessing, scoring, and prioritizing your initiatives using the power of decision intelligence.

Prioriti AI is revolutionizing the way global enterprises approach decision-making. Here's what sets us apart:

Figures 1 - 19 Used with the permission of CartoonStock.com

Acknowledgments

This book was made possible with the encouragement, support, friendships, and family that have guided me through life and my career. I specifically want to acknowledge a few key people.

My family

I want to thank my kids, grandkids, cousins, family members passed but not forgotten, and more who stood beside me while I was endeavoring my career. I also want to express my gratitude to Gloria, who supported me through the challenges of launching my startup.

Steve Mankoff, former Senior Vice President, Siebel Systems and Managing Partner at TDF Ventures.

Steve has been a mentor and friend to me throughout my 25+ year career in Silicon Valley. He has advised and encouraged me to venture into the world of high-tech startups, and I am grateful for his support. I have learned a lot from Steve.

Bruce Cleveland, former CMO at C3.ai, former Senior Vice President, Siebel Systems and former founder of Wildcat Ventures.

Bruce has been advising me and my startup. He was instrumental in helping us define our category using his Traction Gap method. He continues to provide guidance on market fit, thought leadership, and much more than I could ever receive in an MBA class at any top-tier university.

Barbara Walkowski, former General Counsel at Siebel Systems and investor with companies such as C3.ai, senior executive at Snowflake and others.

First, I want to express my gratitude to Barbara for our 25+ years of friendship. She and I worked closely together in the early days of Siebel Systems. From being colleagues to her current role as my advisor, our interactions have always been lively, with plenty of thought-provoking conversations, problem-solving challenges, and more to keep me engaged and alert.

Kimble Ratliff, Southwood Family Holdings.

Thank you, Kim, for your enduring friendship. Your unwavering support during challenging times in my career and life has been invaluable. I couldn't have asked for a better friend and business partner.

To my present and former colleagues.

It has been quite the journey to reach this moment. I have cherished many memorable experiences across the globe, from San Francisco to New York, and from London to Barcelona, Bangalore, and beyond. It has been an incredible ride.

To my friends (from all over the world!)

I want to thank all my friends, from recent to long ago, as well as those who were with me during challenging times. There are too many of you to mention, but I want you to thank you all from Marina del Rey, San Francisco, Tuscaloosa, Atlanta, and Birmingham.